Cognitive Technologies

Managing Editors: D. M. Gabbay J. Siekmann

Srikanta Patnaik

Robot Cognition and Navigation

An Experiment with Mobile Robots

With 95 Figures, and 7 Tables

 Springer

Author:

Srikanta Patnaik
Interscience Institute of Management & Technology
Baranga - Khurda Road
P.O. Kantabada
Bhubaneswar – 752 054
India
srikantapatnaik@interscience.in

Managing Editors:

Prof. Dov M. Gabbay
Augustus De Morgan Professor of Logic
Department of Computer Science, King's College London
Strand, London WC2R 2LS, UK

Prof. Dr. Jörg Siekmann
Forschungsbereich Deduktions- und Multiagentensysteme, DFKI
Stuhlsatzenweg 3, Geb. 43, 66123 Saarbrücken, Germany

Library of Congress Control Number: 2007928310

ACM Computing Classification: I.2, I.4, I.5

ISSN 1611-2482
ISBN 978-3-540-23446-3 Springer Berlin Heidelberg New York

Springer is a part of Springer Science+Business Media
springer.com

© Springer-Verlag Berlin Heidelberg 2007
Printed in Germany

Typesetting: by the author
Production: Integra Software Services Pvt. Ltd., Puducherry, India
Cover Design: KünkelLopka, Heidelberg

Printed on acid-free paper   45/3180/Integra   5 4 3 2 1 0

To my wife

Priti

and my sons

Sritam and Priyam

Preface

This book is meant for graduate and undergraduate students. It contains eighteen chapters, covering robot simulation, experiments and modelling of mobile robots. It starts with the cybernetic view of robot cognition and perception. Chapter 2 discusses map building, which is necessary for the robot to sense and represent its surrounding. Chapter 3 describes path planning in a static environment and a path planning technique using quad-trees. Chapter 4 discusses robot navigation using genetic algorithm techniques.

Chapter 5 briefly covers the robot programming package along with hardware details for the robot experiments. Socket programming and the multithreading concept are also briefly explained for robot programming. Chapter 6 covers a client–server program for robot parameter display. Chapter 7 covers the BotSpeak program and Chap. 8 covers gripper programs. Chapter 9 covers the program for Sonar reading display. Chapter 10 covers the robot program for wandering within the environment and Chap. 11 describes the program for tele-operation. A complete navigation program has been explained in Chap. 12 utilizing all the functions of the earlier chapters.

Chapter 13 describes the techniques for imaging geometry. Image formation and the camera perspective matrix in 3D is covered in the chapter. Chapter 14 describes the program for image capture through the robot's stereo camera. Chapter 15 describes the minimal representation of a 2D line, a 3D line and a 3D plane. The chapter also describes the techniques for reconstruction of 3D points, 3D lines and 3D planes using the Kalman filter. A correspondence problem has also been highlighted in the chapter. The program for 3D perception using the Kalman filter is given in Chap. 16. Chapter 17 describes robot perception for non-planar surfaces. In Chap. 18, a real-time application of the mobile robot is given.

The detailed source codes of various programs mentioned in the book are available in the website (http://www.springer.com/3-540-23446-2).

Acknowledgements

The author gratefully acknowledges the contribution of many people, without whom this publication might not have been possible. First of all, he wishes to thank his undergraduate students doing their final semester project under his guidance in the Electronics and Telecommunication Engineering Department of U.C.E., Burla during the periods 2001–2002, 2002–2003 and 2003–2004. The author would like to convey special thanks to his students Deb Prakash, Priyanka, Sarada Prasanna, Vidyasagar, Amaresh and Bismaya.

The author would like to thank his teacher, Prof. K. C. Mohapatra, who inspired his writing skill, which later enabled him to write this book. He remembers his onetime project supervisor Dr. Rutuparna Panda of the Electronics and Telecommunication Engineering Department of U.C.E., Burla, for his constructive criticism, which helped him develop the habit of checking a thought twice before planning. The author is greatly indebted to his PhD guide, Dr. Amit Konar, and Prof. A. K. Mandal of Jadavpur University, due to whom he conceived the idea of writing a book in this area. The author wants to convey special thanks to his colleagues Dr. Brajamohan Otta and Mr. Rabinarayan Rath, for their personal support and help.

The author is indebted to Prof. Ratikanta Mishra, Vice-Chancellor of Biju Patnaik University of Technology, Rourkela, for his constant help, support and encouragement during the compilation of the book. The author owes deep gratitude to Prof. Sukadev Nanda, Vice-Chancellor of F. M. University, Balasore, for his help and administrative support for the writing and completion of this project.

The author sincerely thanks the All India Council for Technical Education, New Delhi, for the project grant towards the TAPTEC project entitled *Building Cognition for Mobile Robots* for procuring the mobile robot Pioneer DX-2, from ActivMedia Robotics LLC, USA. The author also acknowledges the support of the University Grant Commission's major research project entitled *Machine Learning and Perception Using Cognitive Methods* through which many experiments could be conducted.

The author would like to thank Ingeborg Mayer and Ronan Nugent of Springer for their kind cooperation in connection with writing this book. The author wishes to express his deep gratitude to his parents, who always stood by him throughout his life. Last but not least, the author wishes to thank his wife Priti for her tolerance and forbearance of his indifference to family life and her assistance in many ways for the successful completion of the book. He also conveys special thanks to his sons Sritam and Priyam for their sacrifice and tolerance during the writing of the book.

India, Februay 2007   Srikanta Patnaik

Contents

1   Cybernetic View of Robot Cognition and Perception 1
1.1 Introduction to the Model of Cognition 1
    1.1.1 Various States of Cognition 3
    1.1.2 Cycles of Cognition 5
1.2 Visual Perception 7
    1.2.1 Human Visual System 7
    1.2.2 Vision for Mobile Robots 8
1.3 Visual Recognition 10
    1.3.1 Template Matching 11
    1.3.2 Feature-Based Model 11
    1.3.3 Fourier Model 12
    1.3.4 Structural Model 12
    1.3.5 The Computational Theory of Marr 13
1.4 Machine Learning 13
    1.4.1 Properties and Issues in Machine Learning 13
    1.4.2 Classification of Machine Learning 15
1.5 Soft Computing Tools and Robot Cognition 17
    1.5.1 Modeling Cognition Using ANN 17
    1.5.2 Fuzzy Logic in Robot Cognition 19
    1.5.3 Genetic Algorithms in Robot Cognition 19
1.6 Summary 20

2   Map Building 21
2.1 Introduction 21
2.2 Constructing a 2D World Map 22
    2.2.1 Data Structure for Map Building 22
    2.2.2 Explanation of the Algorithm 26
    2.2.3 An Illustration of Procedure `Traverse Boundary` .. 27
    2.2.4 An Illustration of Procedure `Map Building` 29
    2.2.5 Robot Simulation 31
2.3 Execution of the Map Building Program 33
2.4 Summary 38

3.   Path Planning...**39**
    3.1 Introduction...39
    3.2 Representation of the Robot's Environment..........................39
        3.2.1 GVD Using Cellular Automata...............................40
    3.3 Path Optimization by the Quadtree Approach........................41
        3.3.1 Introduction to the Quadtree.............................41
        3.3.2 Definition...42
        3.3.3 Generation of the Quadtree...............................42
    3.4 Neighbor-Finding Algorithms for the Quadtree 47
    3.5 The A* Algorithm for Selecting the Best Neighbor 52
    3.6 Execution of the Quadtree-Based Path Planner Program 54
    3.7 Summary 58

4   Navigation Using a Genetic Algorithm **59**
    4.1 Introduction...59
    4.2 Genetic Algorithms...60
        4.2.1 Encoding of a Chromosome 61
        4.2.2 Crossover..62
        4.2.3 Mutation 62
        4.2.4 Parameters of a GA 63
        4.2.5 Selection..63
    4.3 Navigation by a Genetic Algorithm................................64
        4.3.1 Formulation of Navigation 64
    4.4 Execution of the GA-Based Navigation Program................67
    4.5 Replanning by Temporal Associative Memory 68
        4.5.1 Introduction to TAM 68
        4.5.2 Encoding and Decoding Process
             in a Temporal Memory....................................70
        4.5.3 An Example in a Semi-dynamic Environment.............71
        4.5.4 Implications of Results....................................74
    4.6 Summary...75

5   Robot Programming Packages **77**
    5.1 Introduction...77
    5.2 Robot Hardware and Software Resources 78
        5.2.1 Components..79
    5.3 ARIA 79
        5.3.1 ARIA Client–Server......................................80
        5.3.2 Robot Communication 84
        5.3.3 Opening the Connection...................................84
        5.3.4 ArRobot..85
        5.3.5 Range Devices...87
        5.3.6 Commands and Actions 88

5.4 Socket Programming 95
    5.4.1 Socket Programming in ARIA 96
5.5 BotSpeak Speech System 98
    5.5.1 Functions 98
5.6 Small Vision System (SVS) 100
    5.6.1 SVS C++ Classes 101
    5.6.2 Parameter Classes 102
    5.6.3 Stereo Image Class 102
    5.6.4 Acquisition Classes 106
5.7 Multithreading 112
5.8 Client Front-End Design Using Java 113
5.9 Summary 113

6   **Robot Parameter Display** **115**
    6.1 Introduction 115
    6.2 Flow Chart and Source Code for Robot Parameter
        Display 115
    6.3 Summary 125

7   **Program for BotSpeak** **127**
    7.1 Introduction 127
    7.2 Flow Chart and Source Code for BotSpeak Program 127
    7.3 Summary 136

8   **Gripper Control Program** **137**
    8.1 Introduction 137
    8.2 Flow Chart and Source Code for Gripper Control
        Program 137
    8.3 Summary 150

9   **Program for Sonar Reading Display** **151**
    9.1 Introduction 151
    9.2 Flow Chart and Source Code for Sonar Reading Display
        on Client 151
    9.3 Summary 161

10   **Program for Wandering Within the Workspace** **163**
    10.1 Introduction 163
    10.2 Algorithm and Source Code for Wandering Within
        the Workspace 163
    10.3 Summary 173

11   Program for Tele-operation..**175**
  11.1 Introduction..175
  11.2 Algorithm and Source Code for Tele-operation 175
  11.3 Summary..188

12   A Complete Program for Autonomous Navigation 189
  12.1 Introduction..189
  12.2 The ImageServer Program. 190
  12.3 The MotionServer Program 192
  12.4 The Navigator Client Program..195
  12.5 Summary..199

13   Imaging Geometry..**201**
  13.1 Introduction..201
  13.2 Necessity for 3D Reconstruction 201
  13.3 Building Perception 202
      13.3.1 Problems of Understanding 3D Objects from 2D
              Imagery 203
      13.3.2 Process of 3D Reconstruction 203
  13.4 Imaging Geometry 205
      13.4.1 Image Formation 205
      13.4.2 Perspective Projection in One Dimension................206
      13.4.3 Perspective Projection in 3D....................................207
  13.5 Global Representation 211
  13.6 Transformation to Global Coordinate System....................217
  13.7 Summary..220

14   Image Capture Program..**221**
  14.1 Introduction..221
  14.2 Algorithm for Image Capture 221
  14.3 Summary..225

15   Building 3D Perception Using a Kalman Filter....................**227**
  15.1 Introduction..227
  15.2 Minimal Representation..227
  15.3 Recursive Kalman Filter 229
  15.4 Experiments and Estimation 231
      15.4.1 Reconstruction of 3D Points 237
      15.4.2 Reconstruction of a 3D Line 242
      15.4.3 Reconstruction of a 3D Plane....................................246
  15.5 Correspondence Problem in 3D Recovery............................249
  15.6 Summary..250

16   Program for 3D Perception..**251**
        16.1 Introduction..251
        16.2 Flow Chart and Source Code for 3D Perception 251
        16.3 Summary...262

17   Perceptions of Non-planar Surfaces.......................................**263**
        17.1 Introduction..263
        17.2 Methods of Edge Detection 263
        17.3 Curve Tracking and Curve Fitting......................................266
        17.4 Program for Curve Detector 270
        17.5 Summary...275

18   Intelligent Garbage Collection..**277**
        18.1 Introduction..277
        18.2 Algorithms and Source Code for Garbage Collection 277
        18.3 Summary...281

References..**283**

Index...**289**

1 Cybernetic View of Robot Cognition
and Perception

1.1 Introduction to the Model of Cognition

The word 'cognition' generally refers to the faculty of mental activities of human beings dealing with abstraction of information from the real world, their representation and storage in memory, as well as automatic recall [Patnaik et al., 2003a]. It includes construction of higher level percepts from low level information or knowledge, which is referred to as perception. It also does the construction of mental imagery from real instances for subsequent usage in recognizing patterns or understanding complex scenes. It includes various behaviors like sensing, reasoning, attention, recognition, learning, planning and task coordination, as well as the control of activities of human beings. Cognitive science is a contemporary field of study that tries to answer questions about the nature of knowledge, its components, development and uses [Matlin, 1984]. Cognitive scientists have the opinion that human thinking involves the manipulation of internal representation of the external world known as the *cognitive model*. Different researchers have investigated various models of human cognition during the last thirty years. Those were basically analytical and experimental psychology, and gradually scientists have tried to implement this knowledge in developing intelligent robots.

A very first model of a robot was developed by dividing the entire task into a few subtasks, namely *sensing, planning, task coordination* and *action,* which was popularly known as the *principle of functional decomposition*. These subtasks were realized on separate modules that together form a chain of information flow from the environment to the actuators, through sensing, planning and task coordination. The primitive model is poor at accommodating major components like perception, map building and world modeling. Subsequently, these functional modules were included in the robot model, by various researchers.

In 1986, Rodney A. Brooks was the first man to use the findings of ethological research, and to design a mobile robot. He published a seminal paper on the *subsumption architecture,* which was fundamentally a different approach in the development of mobile robots [Arbib, 1981]. He developed the *subsumption language* that would allow him to model something analogous to animal behaviors in tight *sense-act* loops using asynchronous finite-state machines. The first type of behavior for a robot was used to avoid obstacles that are too close and moving a little away or else standing still. Secondly, higher level behavior might be to move the robot in a given direction. This behavior would dominate the obstacle-avoidance behavior by suppressing its output to the actuators unless an object comes too close. The higher levels subsumed the lower levels, and therefore the name of the architecture was *subsumption architecture.* They were able to develop a robot, using simple sonar or infrared sensors that could wander around a laboratory for hours without colliding into objects or moving people. After this development, Brooks and his colleagues developed highly mobile robots, i.e. mobots, both wheeled and legged, which could chase moving objects or people and run or hide from light. Further, they can negotiate a cluttered landscape which might be found in a rugged outdoor environment.

During the 1990s, there were many developments such as HERBERT: a soda-can-collecting robot [Connell, 1990]; GENGHIS: a robot that learned to walk [Maes & Brooks, 1990; Brooks, 1989]; TOTO: a hallway-navigating robot [Matric, 1992]; and POLLY: a tour-guide robot [Horswill, 1993]. The idea was to build up capability in the robot through behaviors that run in parallel to achieving possible alternative goals. The behavior in these robots was able to execute various actions on a priority basis or to achieve various goals within the cycle time. After this development, the world model was distributed among the types of behaviors with only relevant parts of the model being processed for each behavior. The generation of simple plans for path planning and the compilation of the result of actions could be done before run time by using these types of behaviors.

Today, researchers are trying to develop intelligent machines after a careful and meticulous review of human cognition, soft computing tools and techniques. But there are still open problems in these areas of machine learning and perception, which are being investigated using many alternative approaches. This research work mainly aims at studying various techniques of perception and learning, using the cognitive model, and their applications in mobile robots. Detailed programs have been provided in the respective chapters.

1.1.1 Various States of Cognition

Let us introduce a model of cognition that includes seven mental states, namely *sensing and acquisition, reasoning, attention, recognition, learning, planning, action and coordination* and their transitions along with *cognitive memory,* i.e. *LTM* and *STM,* as shown in Fig. 1.1. There are three cycles embedded in the model, namely the *acquisition cycle, the perception cycle* and *the learning and coordination cycle,* which describe the concurrent transition of various states.

The acquisition cycle consists of two states namely *sensing and attention* along with *short term memory* (STM) and *long term memory* (LTM), whereas the *perception cycle* consists of three states namely *reasoning, attention and recognition* along with LTM. The *learning and coordination* cycle consists of LTM along with three states namely, *learning, planning* and *action.* The various states are explained as follows.

Sensing and acquisition: Sensing in engineering science refers to reception and transformation of signals into a measurable form, which has a wider perspective in cognitive science. It includes preprocessing and

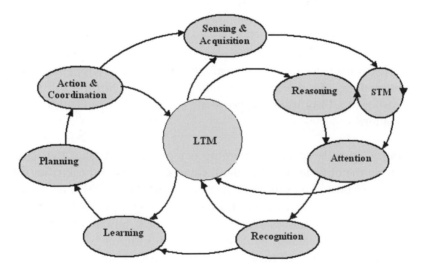

Fig. 1.1. Three cycles namely acquisition, perception and learning and coordination, with their states in the model of cognition. LTM = Long Term Memory; STM = Short Term Memory

extraction of features from the sensed data, along with stored knowledge in LTM. For example, visual information on reception is filtered from undesirable noise, and the elementary features like size, shape, color are extracted and stored in STM [Borenstain, 1996].

Reasoning: Generally this state constructs high level knowledge from acquired information of relatively lower level and organizes it in structural form for efficient access [Bharick, 1984]. The process of reasoning analyses the semantic or meaningful behavior of the low level knowledge and their association [Chang, 1986]. It can be modeled by a number of techniques such as commonsense reasoning, causal reasoning, non-monotonic reasoning, default reasoning, fuzzy reasoning, spatial and temporal reasoning, and meta-level reasoning [Popovic et al., 1994].

Attention: This is responsible for the processing of a certain part of the information more extensively, while the remaining part is neglected or suppressed. Generally, it is task-specific visual processing which is adopted by animal visual systems [Matlin, 1984]. For instance, finding out the area of interest in a scene autonomously is an act of attention.

Recognition: This involves identifying a complex arrangement of sensory stimuli such as a letter of the alphabet or a human face from a complex scene [Murphy et al., 1998]. For example, when a person recognizes a pattern or an object from a large scene, his sensory-organs process, transform and organize the raw data received by the sensory receptors. Then the process compares the acquired data from STM with the information stored earlier in LTM through appropriate reasoning for recognition of the sensed pattern.

Learning: Generally speaking, learning is a process that takes the sensory stimuli from the outside world in the form of examples and classifies these things without providing any explicit rules [Winston, 1975]. For instance, a child cannot distinguish between a cat and a dog. But as he grows, he can do so, based on numerous examples of each animal given to him. Learning involves a teacher, who helps to classify things by correcting the mistake of the learner each time. In machine learning, a program takes the place of a teacher, discovering the mistakes of the learner. Numerous methods and techniques of learning have been developed and classified as supervised, unsupervised and reinforcement learning [Baldi, 1995; Carpenter et al., 1987; Lee et al., 1997].

Planning: The state of planning engages itself to determine the steps of action involved in deriving the required goal state from known initial states of the problem. The main task is to identify the appropriate piece of knowledge derived from LTM at a given instance of time [McDermott et al., 1984]. Then planning executes this task through matching the problem states with its perceptual model.

Action and coordination: This state determines the control commands for various actuators to execute the schedule of the action plan of a given problem, which is carried out through a process of supervised learning [Maes & Brooks, 1990]. The state also coordinates between various desired actions and the input stimuli.

Cognitive memory: Sensory information is stored in the human brain at closely linked neuron cells. Information in some cells may be preserved only for a short duration, which is referred to as short term memory (STM). Further, there are cells in the human brain that can hold information for quite a long time, which is called long term memory (LTM). STM and LTM could also be of two basic varieties, namely *iconic memory* and *echoic memory*. Iconic memory can store visual information whereas the echoic memory deals with audio information. These two types of memories together are generally called *sensory memory*. Tulving alternatively classified human memory into three classes, namely episodic, semantic and procedural memory [Tulving, 1987]. Episodic memory saves the facts as they happen; semantic memory constructs knowledge in structural form, whereas procedural memory helps in taking decisions for actions.

1.1.2 Cycles of Cognition

Acquisition cycle: The task of the acquisition cycle is to store the information temporarily in STM after sensing the information through various sensory organs. Then it compares the response of the STM with already acquired and permanently stored information in LTM. The process of representation of the information for storage and retrieval from LTM is a critical job, which is known as *knowledge representation.* It is not yet known how human beings store, retrieve and use the information from LTM.

Perception cycle: This is a cycle or a process that uses the previously stored knowledge in LTM to gather and interpret the stimuli registered by the sensory organs through the acquisition cycle [Gardener, 1985]. Three relevant states of perception are *reasoning, attention* and *recognition,* and are generally carried out by a process of unsupervised learning. Here, we can say that the learning is unsupervised, since such refinement of knowledge is an autonomous process and requires no trainer for its adaptation. Therefore, this cycle does not have "Learning" as an exclusive state. It is used mainly for feature extraction, image matching and robot world modeling. We will discuss human perception in detail in the next section, with applications.

Learning and coordination cycle: Once the environment is perceived and stored in LTM in a suitable format (data structure), the autonomous system utilizes various states namely *learning, planning* and *action and coordination* [Caelli & Bischob, 1997]. These three states taken together are called the *Learning and Coordination Cycle*, which is utilized by the robot to plan its action or movement in the environment.

Cognition, being an interdisciplinary area, has drawn the attention of researchers of diverse interest. Psychologists study the behavioral aspects of cognition and they have constructed a conceptual model that resembles the behavior of cognition with respect to biological phenomena. On the other hand, the engineering community makes an attempt to realize such behavior on an *intelligent agent* by employing AI and soft computing tools. The robot as an intelligent agent receives sensory signals from its environment and acts on it through its actuators as well as sensors to execute physical tasks.

This book covers techniques for feature extraction, image matching, machine learning and navigation using the cognitive method. A mobile robot senses the world around it through different transducers, such as ultrasonic sensors, laser range-finders, drives and encoders, tactile sensors, and mono or stereo cameras. The sensory information obtained by a robot is generally contaminated with various forms of noise. For instance, ultrasonic sensors and laser range-finders sometimes generate false signals, and as a consequence determination of the direction of an obstacle becomes difficult. The acquisition cycle filters the contaminated noise and transfers the noise-free information to the LTM. The perception cycle constructs new knowledge of the robot's environment from the noise-free sensory information. The learning and coordination cycle executes various tasks assigned to it. These three cycles, along with their states, are utilized in modeling various techniques. The proposed work borrows the ideas from the model of cognition and implements them through various soft computing tools. The subsequent sections discuss issues like machine learning and perception in detail.

1.2 Visual Perception

Vision is the most powerful sense organ of human beings and it is also the key sensory device for a mobile robot. So far, not much progress has yet been achieved in visual perception of a mobile robot due to limitations in hardware and software. Visual processing requires specialized hardware and cameras, which are quite large to fit into the mobile robot. Secondly, traditional software for vision processing is very poor in quality, because it requires complete analysis of the entire scene even to recognize a minute object. Further, detection of an obstacle in front of the robot using stereovision takes a longer time, which is not at all permissible for online navigation.

As there is a shift in paradigm towards the behavioral model, researchers have started examining animal models for both motor control and perception. The research findings reveal that the frog uses simple visual motion detection to catch flying prey and bees depend on specific aspects of the color spectrum for their search. Psychological study indicates that the human visual system supports very simple behavior. Low resolution peripheral vision is used to watch for indications of motion, for instance collision with looming objects, whereas the high resolution fovea is used to gather information for reasoning about an object. Human vision does not perceive everything in all its color, motion and temporal dimensions at one time but direct attention is given to a very narrow portion of the visual field based on the task they are performing. As a result of this study, the paradigm in vision is shifted to a philosophy where perception exists to support the behavior of robots [Murphy et al., 1998].

Dickmann studied two major principles of perception during the 1980s. The first one was about the evolving process and internal representation of the world, which is known as Schopenhauer's idea of perception, and the second one was Kant's theory of the *true reality of perception*. With these two principles, Dickmann could represent various systems, including real time constraint using the notion of space and time [Zavidovique, 2002]. Human beings acquire knowledge of their surroundings unconsciously during their first years of crawling, then walking and reacting. Let us discuss a little more, the physiology and anatomy of the human visual system, which may help in understanding robot vision.

1.2.1 Human Visual System

The human visual system converts energy in the visible spectrum into action potentials in the optic nerves. The wavelength of visible light ranges

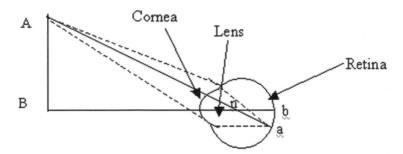

Fig. 1.2. A schematic diagram of human vision. n = nodal point, AnB and anb are similar triangles. In this diagram, the nodal point is 15 mm from the retina. All refraction is assumed to take place at the surface of the cornea, 5 mm from the nodal point. The dotted lines represent rays of light diverging from A and refracted at the cornea so that they are focused on the retina at 'a'

from approximately 390 nm to 720 nm. In the human eye, light is actually refracted at the anterior surface of the cornea and at the anterior and posterior surface of the lens. The process of refraction is shown schematically in Fig. 1.2.

The images of objects in the environment are focused on the retina. The light rays striking the retina generate potential changes that initiate action potentials on photosensitive compounds such as rods and cones. When the compound absorbs light, the structure changes and triggers a sequence of events that initiate neural activity. The human eye contains about 130 million rods and approximately 8 million cones. Rods are monochrome receptors of light, and cones are sensitive to color, which are different parts of the spectrum [Seculer et al., 1990]. The distribution of rods and cones in the retina is very irregular. The detailed spatial representation of the retina is in the form of electrical responses transmitted to the lateral geniculate nucleus (LGN) via retinal ganglion cells (RGC). Subsequently, LGN projects a similar point-to-point representation on the visual cortex, where these electrical responses produce the sensation of vision.

1.2.2 Vision for Mobile Robots

Vision is the fundamental part of perception in intelligent robots, the way it is for humans. The perception objective depends on three basic system qualities, namely *rapidity, compactness* and *robustness*. Active vision is the theoretical analysis of the vision process originated by Aloimonos et al. [Aloimonos, 1987; Fermuller & Aloimonos, 1993] to optimize the 3D reconstruction; and animate vision [Ballard, 1991], which is based on human perception analysis. Perception is an essential and most useful process for a mobile robot. To operate the mobile robot in unknown and unstructured environments, the robot must be able to perceive its environment sufficiently so as to operate safely in its environment. It is clear from study of the animal visual system that the animals concentrate on a specific part of the image received through their visual system, or in other words animal visual system does the task-specific visual processing. This philosophy has been borrowed to develop a task-oriented approach for sensing, planning and control.

In the realm of autonomous control, let us briefly mention visually guided control systems and the role of computer vision in autonomously guided robot systems. Hashimoto has introduced a closed-loop control system for visually guided manipulators [Hashimoto, 1999]. Visual information about tasks and the environment is essential for robots to execute flexible and autonomous tasks. A typical task of autonomous manipulation is to track a moving object with the robot hand, based on information from a vision sensor. To carry out this job, Hashimoto introduced a feedback loop with vision sensor, which can track a moving obstacle (Khepera robot) that moves on the floor. Sullivan et al. [Sullivan, 1999] introduced a system that tracks a moving, deformable object in the workspace of a robotic arm fitted with a camera. For their experiment, they used a figure-ground approach for object detection and identification. The figure-ground methodology allows pixels to be identified as objects or background pixels, a distinction which is useful during the initial placement of the active deformable model.

The evolution of machine perception has taken place during the last few decades [Zavidovique, 2002; Merlo et al., 1987]. The first system for signal analysis and pattern recognition was designed on a commercial computer in which a special-purpose interface was built for data acquisition using standard cameras and microphones. At that time, serious limitations were faced in signal transfer rate and core memory. Both sound and image preprocessing was developed during the 1970s using ad hoc hardware and suitable algorithms. Later, to improve the machine perception strategies, the link between sensing and processing was operated in a closed loop to obtain so-called *active perception*. *Active perception* is the study of perception strategies including sensor and signal processing cooperation, to achieve knowledge about the environment [Merlo et al., 1987]. Both high level and low level image processing need to consider perceptual information in order to reduce uncertainty.

Initially, the aim of the robot vision designer was to build the simplest possible system that was necessary to solve a given task and use its performance to improve its architecture. Horswill has developed a low cost vision system for navigation called the POLLY System, where the development team has used active, purposeful and task-based vision [Horswill, 1993], which computes the specific information needed to solve specific tasks. Murphy has introduced another model of sensing organization called action-oriented perception for a mobile robot, with multiple sensors performing locomotive tasks [Murphy, 1998], and work on robot vision and perception is still continuing as a major research topic.

1.3 Visual Recognition

As mentioned earlier, recognition is the important component of perception. Human beings are able to perceive and move around in a dynamic world without any difficulty but robot vision requires a large amount of computing resources and background knowledge for a very limited and static environment. There are many hypotheses of representation for various shape recognition techniques [Caelli & Bischob, 1997]. The representation contains information about the shape and other properties such as color, dimensions and temporal information, including a label, i.e. a name of the object. The objective is to retrieve the label correctly during the recognition process. Representations are stored in LTM as a set of separable symbolic objects or class of objects.

Recently developed models do not use a representation that is a direct replica of the retinal stimulation. Rather, they introduce the representation which deals with invariant properties of different objects in various positions, sizes, rotations, and even under different lighting conditions. During recognition, the captured image corresponding to an unknown object is converted to the same format and representations, which provides the best match using the same form of similarity measure. Each theory may have different assumptions regarding various parameters, such as:

- Type of representation, i.e. feature space, predicates, graph, etc.
- The number of representations per object, i.e. one 3D representation or multiple 2D representations from different viewing positions
- The number of classes for mapping into representations
- Inclusion of spatial relationships between objects and their component parts
- The amount and type of preprocessing given to the initial retinal image matching algorithm.

The primary issues in the visual recognition process are representations and search, which means how to develop an appropriate representation for the objects and then how to search them efficiently for a match at the time of recognition. Here are some representations used in traditional theories.

1.3.1 Template Matching

Template matching is the simplest form of representation in which a replica of the retinal stimulation pattern projected by a shape is stored in LTM. The recognition process compares all stored object templates with the input array by selecting the best match based on the ratio of matching to non-matching objects [Briscoe, 1997].

There are many problems with this method which prevent recognition, such as: (i) partial matches can give false results, for example comparing 'O' with 'Q'; (ii) any change in the distance, location or orientation of the input object in relation to the corresponding stored object will produce a different pattern; and (iii) any occlusion, shadow or other distortion of the input object may also produce inaccurate matching.

Some systems, for example, attempt to compensate for these problems by storing multiple templates, each recorded at various displacements, rotations and sizes. However, the combinatorics of the transformations usually prove to be cumbersome. The option of rotating, displacing or scaling of the input pattern to a canonical form before matching is also not feasible, as the required transformations cannot be known until the object is recognized. But the major limitation of this representation is that it is only appropriate for an object recorded in isolation. The template models are not useful at all if multiple objects are present in a scene, because the method is unable to determine which parts belong to which object.

1.3.2 Feature-Based Model

Instead of storing templates for entire shapes, the feature-based model utilizes a series of *feature detectors*. Generally the features included are of a geometric type such as vertical and horizontal lines, curves and angles. Feature detectors may be used either at every position in the input array, or may be used for the global image. In case of multiple feature detectors, the degree of matching is estimated for the target feature with respect to each section of the input array. The levels of activation for each feature may be summed up across the input array by providing a set of numbers for each feature. This list of numbers in the form of a vector of weights for different features is used as the stored representation of the object. The objective is to define the shape of the object with invariant features, which are independent of locations. The process of recognition consists of finding the best match between the stored representations and the levels of activation of the feature detectors in the input image.

1.3.3 Fourier Model

In the Fourier model, a two-dimensional input array is subjected to a spatial Fourier analysis. In this model, the original array is decomposed into a set of spatial frequency components of various orientations and frequencies in the form of sinusoidal waveforms. The amplitude and phase are both recorded for the spectrum of spatial frequencies and angles. Thus the original image is represented as the sum of the spatial frequency components and this transform retains all the details of the original image. The feature of this model is that it gives no restriction on angles, frequencies and no computational problems, such as *aliasing*. Even though the amplitude spectrum contains shape information and the phase spectrum contains position information, there is no method available for combining this information in order to locate a particular object at a particular location.

Each shape is stored in the memory in the form of its Fourier transform and its recognition is done by matching this with a similarly transformed input image. This model separates information about sharp edges and small details from other information pertaining to gross overall shape. Techniques such as edge detectors and convolutions may be used to extract these different details of the original image. The advantage of this model is that it can also match blurred edges, wiggly lines and other slightly distorted images.

1.3.4 Structural Model

The structural model contains information about the relative positions and relationships between parts of an object. This structural description is stored in memory in the form of a data structure such as a list or tree or graph of predicates. The representation is often depicted as a graph, where nodes correspond to the parts or the properties, and the edges correspond to the spatial relations [Minsky, 1975]. The advantage of structural representation is that it factors apart the information in a scene without losing any part of it. This model enables us to represent the object with the help of a list of labels and also their relative position and orientation with respect to the human observer. Various spatial reasoning operations may be performed by specifying the shape, location, orientation and spatial relationship of one set of objects with other objects in another set. The recognition process can be improved by including statistical and logical operations. The use of structural descriptions appears to be preferred because of computational convenience.

1.3.5 The Computational Theory of Marr

The work of David Marr [Marr, 1982] is one of the best examples of the computational approach to the recognition problem, which is the most influential contemporary model of 3D shape recognition. Marr introduced the need to determine edges of an object and constructed a 2½ D model, which carries more information than 2D but less than a 3D image. Thus an approximate guess about the 3D object can be framed from its 2½ D images.

1.4 Machine Learning

Since the invention of the computer there was always the question of how to make them learn. If we could understand how to program them to learn, i.e. to improve automatically with experience, it would have been a great achievement. A successful understanding of how to make them learn would open up many new uses of computers and new levels of competence and customization. Further, a detailed understanding of machine learning might lead to further investigation of human learning ability and disabilities. Machine learning algorithms have been investigated by a number of researchers. These are effective for certain types of learning tasks and as a result a theoretical understanding of learning started to emerge. Broadly speaking learning means any computer program that improves its performance for some tasks through experience.

1.4.1 Properties and Issues in Machine Learning

The formal definition of learning is: *A computer program is said to learn from experience E with respect to some class of tasks T and performance measure P, if its performance at tasks in T as measured by P improves with experience E* [Mitchell, 1997]. To give a specific example of learning, let us consider an example of an autonomous driven vehicle. The machine learning method has been used to train a computer-trained vehicle to steer correctly when driving on a variety of roads. For instance, the

ALVINN system [Pomerleau et al., 1989] has used its learned strategies to drive unassisted at 70 miles/hour for 90 minutes on public highways among other cars. Similar techniques have possible applications in many sensor-based control systems. This learning problem of autonomous driven systems can be formally defined as

Task T: driving on public highways using vision sensors.
Performance measure P: average distance traveled before an error occurs as judged by a human observer.
Training experience E: a sequence of images and steering commands recorded while observing a human driver.

There are various factors to be considered while designing a learning system, which are given as follows.

Choosing the training experience: The type of training experience available can have a significant impact on the success or failure of the learner. There are a few key attributes to contribute to the success of the learner. The first key is whether the training experience provides direct or indirect feedback regarding the choices made by the performance system. The second attribute of the training experience is the degree to which the learner controls the sequence of training examples. The next attribute is how well it represents the distribution of examples over which the final system performance P must be measured. Generally speaking, learning is most reliable when the training examples follow a distribution similar to that of future test examples.

Choosing the target function: This is to determine exactly what type of knowledge will be learned and how this will be used by the performance program. In other words, the task is to discover an operational description of the ideal target function or an approximation to the target function. Hence, the process of learning the target function is often called the function approximation.

Choosing the representation of a target function: A representation of the target function has to be described for the learning program to learn. In general the choice of representation involves a crucial trade-off. On one hand, a very expressive representation has to be picked up in order to obtain an approximate representation function as close as possible to the ideal target function. On the other hand, the program will require more training data for the more expressive representation. Therefore, it is advisable to choose an alternative representation, which can accommodate a broad state of training data and function.

Choosing a function approximation algorithm: In order to learn the target function, a set of befitting training examples is required. For instance, in the case of robot navigation a set of sensory readings and robot movement forms a set of training patterns for learning. After the derivation of the training examples from the training experience available to the learner, the weights have to be adjusted to best fit these training examples.

One of the useful perspectives on machine learning is that it involves searching every large space of possible hypotheses to determine one that best fits the observed data and any prior knowledge held by the learner. The machine learning algorithm has proven itself useful in a number of application areas, such as:

- poorly understood domains where humans might not have required knowledge to develop effective algorithms, for example human face recognition from images;
- domains where the program must dynamically adapt to changing conditions, for example controlling manufacturing process under changing supply stock;
- they are especially useful in data mining problems where large databases may contain valuable implicit regularities that can be discovered automatically.

1.4.2 Classification of Machine Learning

Machine learning can be broadly classified into three categories: (i) *supervised learning*, (ii) *unsupervised learning* and (iii) *reinforcement learning*. *Supervised learning* requires a trainer, who supplies the input–output training instances. The learning system adapts its parameters using some algorithms to generate the desired output patterns from a given input pattern. But, in the absence of trainers, the desired output for a given input instance is not known, and consequently the learner has to adapt its parameters autonomously. This type of learning is termed *unsupervised learning*. There is a third type of learning, known as *reinforcement learning*, which bridges the gap between supervised and unsupervised categories. In reinforcement learning, the learner does not explicitly know the input–output instances, but it receives some form of feedback from its environment. The feedback signals help the learner to decide whether its action on the environment is rewarding or punishable. The learner thus adapts its parameters based on the states of its actions, i.e. rewarding or punishable. Recently, a fourth category of learning has emerged from the disciplines of knowledge engineering, which is known as *inductive logic programming* [Konar, 2000].

Supervised learning: In supervised learning a trainer submits the input–output exemplary patterns and the learner has to adjust the parameters of the system autonomously, so that it can yield the correct output pattern when excited with one of the given input patterns. Inductive learning [Michalski, 1983] is a special class of the supervised learning technique, where, given a set of $\{x_i, f(x_i)\}$ pairs, a hypothesis $h(x_i)$ is determined such that $h(x_i) \approx f(x_i), \forall i$. This demonstrates that a number of training instances are required to form a concept in inductive learning.

Unsupervised learning: Unsupervised learning employs no trainer and the learner has to construct concepts by experimenting on the environment. The environment responds but fails to identify which ones are rewarding and which ones are punishable activities. This is because of the fact that the goals or the outputs of the training instances are unknown. Therefore, the environment cannot measure the status of the activities of the learner with respect to the goals. One of the simplest ways to construct a concept by unsupervised learning is through experiments. For example, suppose a child throws a ball to the wall; the ball bounces and returns to the child. After performing this experiment a number of times, the child learns the 'principle of bouncing', which is an example of unsupervised learning.

An intelligent system should be able to learn quickly from large amounts of data [Kasabov, 1998]. It is also stated that the machine should adapt in real time and in an online mode as new data is encountered. It should be memory based and possess data and exemplary storage and retrieval capacities. Secondly, the system should be able to learn and improve through active interaction with the user and the environment. But not much progress has been achieved for this learning to date.

Reinforcement learning: In reinforcement learning, the learner adapts its parameter by determining the status, i.e. reward or punishment of the feedback signals from its environment. The simplest form of reinforcement learning is adopted in learning automata. Currently Q learning and temporal difference learning have been devised based on the reward/punishment status of the feedback signals.

1.5 Soft Computing Tools and Robot Cognition

A collection of tools shared by *artificial neural nets, fuzzy logic, genetic algorithms, belief calculus*, and some aspects of *inductive logic programming* are known as *soft computing tools*. These tools are used independently as well as jointly depending on the type of the domain of application [Jain, 1999]. According to Zadeh, soft computing is "an emerging approach for computing, which parallels the remarkable ability of the human mind to reason and learn in an environment of uncertainty and imprecision" [Zadeh, 1983]. The scope of these tools in modeling robot cognition is outlined below.

1.5.1 Modeling Cognition Using ANN

As we know the goal of cognitive modeling is the development of algorithms that require machines to perform cognitive tasks at which humans are presently better [Haykins, 1999]. A cognitive system must be capable of (i) sensing the external environment and storing it in the form of knowledge, (ii) applying the knowledge stored to solve problems, and (iii) acquiring new knowledge through experience. To perform this task the machine needs representation, reasoning and learning.

Machine learning may involve two different kinds of information processing, i.e. inductive and deductive. In inductive processing, generally patterns and rules are determined from raw data and experience; whereas in deductive processing general rules are used to determine specific facts. Similarity-based learning uses induction whereas the proof of a theorem is a deduction from known axioms and other existing theorems. Both induction and deduction processing can be used for explanation-based learning. The importance of knowledge bases and difficulties experienced in learning has led to the development of various methods for supplementing knowledge bases. Specifically, if there are experts in a given field, it is usually easier to obtain the compiled experience of the experts, rather than use direct experience. In fact, this is the idea behind the development of neural networks as a cognitive model. Let us compare neural networks with the cognitive model with respect to three aspects, namely level of explanation, style of processing and representation of structure.

Level of explanation: In traditional machine intelligence, the emphasis is given to building symbolic representations of the problem. It assumes the existence of a mental representation and it models cognition as the sequential processing of symbolic representations [Newell et al., 1972]. On the other hand, a neural network emphasizes the development of parallel distribution processing models. These models assume that information processing takes place through the interaction of a large number of neurons, each of which sends excitatory and inhibitory signals to other neurons in the network [Rumelhart et al., 1986]. Moreover, neural networks give more emphasis to the neurobiological explanation of cognitive phenomenon.

Style of processing: In traditional machine intelligence, processing is sequential as in typical computer programming. Even when there is no predetermined order, the operations are performed in a stepwise manner. On the other hand, neural networks process the information in parallel and provide flexibility about the structure of the source. Moreover, parallelism may be massive which gives neural networks a remarkable form of robustness. With the computation spread over many neurons, it usually does not matter much if the states of some neurons in the network deviate from their expected values. Noisy or incomplete inputs may still be recognized, and may be able to function satisfactorily and therefore learning does not have to be perfect. Performance of the network degrades within a certain range. The network is made even more robust by virtue of coarse coding, where each feature is spread over several neurons [Hinton, 1981].

Representation of structure: In traditional machine intelligence, representation is done through a language of thought, which possesses a quasi-linguistic structure. These are generally complex to build in a systematic fashion from simple symbols. In contrast, the nature and structure of representation is very crucial in neural networks. For implementation of cognitive tasks, a neural network emphasizes the approach to building a structure connectionist model that integrates them. As a result a neural network combines the desirable features of adaptability, robustness and uniformity with representation and inference [Feldman, 1992; Waltz, 1997].

The ANNs adjust the weights of the neurons between different layers during the adaptation cycle. The adaptation cycle is required for updating various parameters of the network, until a state of equilibrium is reached, following which the parameters no longer change. ANNs support both supervised and unsupervised learning as mentioned earlier. The supervised learning algorithms realized with ANN have been successfully applied in control, automation, robotics and computer vision [Narendra et al., 1990]. On the other hand, unsupervised learning algorithms built with ANNs have been applied in *scheduling, knowledge acquisition* [Buchanan, 1993], *planning* [McDermott et al., 1984] and *analog to digital conversion of data* [Sympson, 1988].

1.5.2 Fuzzy Logic in Robot Cognition

Fuzzy logic deals with fuzzy sets and logical connectives for modeling the human-like reasoning problems of the real world. A fuzzy set, unlike conventional sets, includes all elements of the universal set of the domain with varying membership values in the interval [0,1]. It may be noted that a conventional set contains its members with a membership value equal to one and disregards other elements of the universal set with a zero membership value. The most common operators applied to fuzzy sets are AND (minimum), OR (maximum) and negation (complementation), where AND and OR have binary arguments, while negation has a unary argument. The logic of *Fuzzy Set Theory* was proposed by Zadeh [Zadeh, 1983], who introduced the concept of system theory, and later extended it for approximate reasoning in expert systems. Other pioneering research contributions on *Fuzzy Logic* include the work of Tanaka in stability analysis of control systems [Tanaka, 1995], Mamdani in *cement kiln control* [Mamdani, 1977], Kosko [Kosko, 1994] and Pedrycz [Pedrycz, 1995] in *Fuzzy Neural Nets*, Bezdek in *Pattern Classification* [Bezdek, 1991], and Zimmerman [Zimmerman, 1991] and Yager [Yager, 1983] in *Fuzzy Tools and Techniques*.

Fuzzy logic has become a popular tool for robot cognition in recent years [Saffioti, 1997]. Given the uncertain and incomplete information about the environment available to the autonomous robot, fuzzy rules provide an attractive means for mapping ambiguous sensor data to appropriate information in real time. The methodology of fuzzy logic appears very useful when the processes are too complex for analysis by conventional quantitative techniques or when the available sources of information are interpreted qualitatively, inexactly, or uncertainly, which is the case with mobile robots. However, fuzzy logic parameters are usually determined by domain experts using a trial and error method. Also, as the number of input variables increases, in the case of mobile robots, the number of rules increases exponentially, and this creates much difficulty in determining a large number of rules.

1.5.3 Genetic Algorithms in Robot Cognition

Genetic algorithms (GAs) are stochastic in nature, and mimic the natural process of biological evolution [Rich et al., 1996]. This algorithm borrows the principle of *Darwinism*, which rests on the fundamental belief of the *survival of the fittest* in the process of natural selection of species. GAs find extensive applications in intelligent search, machine learning and optimization problems. The problem states in a GA are denoted by chromosomes, which are usually represented by binary strings. The most common operators used in GAs are crossover and mutation. The evolutionary cycle in a GA consists of the following three sequential steps [Michalewicz, 1986];

(i)   generation of a population (problem states represented by chromosomes)

(ii)   selection of better candidate states from the generated population

(iii) genetic evolution through crossover followed by mutation.

In step (i) a few initial problem states are first identified and in step (ii) a fixed number of better candidate states are selected from the generated population. Step (iii) evolves a new generation through the process of crossover and mutation. These steps are repeated a finite number of times to obtain the solution for the given problem.

GAs have been successfully applied to solve a variety of theoretical and practical problems by imitating the underlying processes of evolution, such as selection, recombination, and mutation. The GA-based approach is a well-accepted technique for enabling systems to adapt to different control tasks [Filho, 1994]. But, it is not feasible for a simple GA to learn online and adapt in real time. The situation is worsened by the fact that most GA methods developed so far assume that the solution space is fixed, thus preventing them from being used in real-time applications [Michalewicz, 1986].

1.6 Summary

This chapter briefly highlights the development of various models of mobile robots and the paradigm shift towards the model of cognition. A model of cognition has been introduced here, which will be realized for various tasks of simulated robots as well as for the mobile robots in subsequent chapters. The chapter defines the term 'cognition' along with its embedded cycles namely the acquisition cycle, perception cycle, learning and coordination cycle and their associated states. A brief review of visual perception, visual recognition and machine learning has been given in subsequent sections. The application of various soft computing tools like fuzzy logic, genetic algorithms and artificial neural networks for robot cognition has also been outlined.

2 Map Building

2.1 Introduction

The phrase *map building* [Patnaik et al., 1998] refers to the construction of a map of the work space autonomously by the robot, which enables the robot to plan the optimal path to the goal. Map building helps the mobile robot to become conversant with the world around it. The information about the neighborhood world of the robot is thus required to be encoded in the form of a knowledge base. For the purpose of navigational planning, a mobile robot must acquire knowledge about its environment. This chapter demonstrates the scope of map building of a mobile robot of its workspace.

It is evident from the discussion in the last chapter that the *acquisition of knowledge* is a pertinent factor in the process of building perception. Human beings can acquire knowledge from their environment through a process of automated learning. Machines too can acquire knowledge by sensing and integrating consistent sensory information. A number of techniques are prevalent for automated acquisition of knowledge. The most common among them are unsupervised and reinforcement learning techniques. In an unsupervised learning scheme, the system updates its parameter by the analyzing consistency of the incoming sensory information. Reinforcement learning, on the other hand, employs a recursive learning rule that adopts the parameters of the systems, until convergence occurs, following which the parameters become time invariant. The chapter includes a technique for constructing a 2D world map by a point mass robot with its program written in C++.

2.2 Constructing a 2D World Map

It is assumed that the height of the robot is less than that of the obstacles within the workspace. In fact, most of the navigational problems for robots are confined to two-dimensional environments. The algorithm for map building in a 2D environment is given below.

There exist two different types of algorithms for automated map building. The first one refers to landmark-based map building [Taylor et al., 1998]and the second one is metric-based map building [Asada, 1990; Elfes, 1987; Pagac et al., 1998]. Offline map building is discussed here utilizing a metric-based approach [Patnaik et al., 1998]. Further, to maintain the order in the traversal of the robot around the obstacles, a directed search is preferred here. The *depth-first search,* which is a directed search technique, is being utilized here.

2.2.1 Data Structure for Map Building

Let us consider a circular mobile robot (shown in Fig. 2.1) that can orient itself in any of the following eight directions: north (N), north-east (NE), east (E), south-east (SE), south (S), south-west (SW), west (W) and north-west (NW).

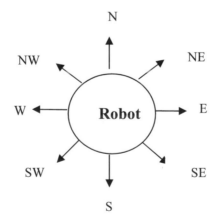

Fig. 2.1. The representation of a circular robot with eight ultrasonic sensors around it in eight geographical directions

In the *depth-first* algorithm, the following strategy is used for traversal within the workspace. The detail explanation is given in the next section:

If there is an obstacle in N
  Then move to the nearest obstacle in N
If there is an obstacle in NE
  Then move to the nearest obstacle in NE
If there is an obstacle in E
  Then move to the nearest obstacle in E
....
....
If there is an obstacle in NW
  Then move to the nearest obstacle in NW

If the above steps are executed recursively, then the robot would have a tendency to move to the north so long as there is an obstacle in the north, else it moves north-east. The process is thus continued until all the obstacles are visited. Another point needs to be noted here, that after moving to an obstacle, the robot should move around it to identify the boundary of the obstacles. Thus when all the obstacles are visited a map representing the boundary of all obstacles will be created. This map is hereafter referred to as the *2D world map* of the robot. Two procedures are given below, i.e. Map Building and Traverse Boundary. In the procedure Map Building a linked list structure is used with four fields. The first two fields, x_i, y_i denote the coordinate of the point visited by the robot. The third field points to the structures containing the obstacle to be visited next, and the fourth field denotes the pointer to the next point to be visited on the same obstacle. A schematic diagram depicting the data structure is presented in Fig. 2.2. Another data structure is used in procedure Map Building for acquiring the boundary points visited around an obstacle. This structure has three fields, the first two correspond to the x_i, y_i coordinate of one visited point on the obstacle i, while the third field is a pointer which corresponds to the next point on obstacle i. A schematic diagram for this pointer definition is presented in Fig. 2.3.

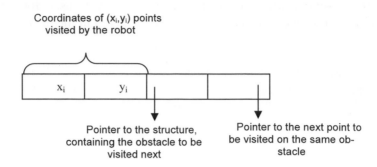

Fig. 2.2. Definition of one structure with two pointers, used for acquiring the list of visited obstacles

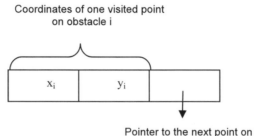

Fig. 2.3. Definition of another structure with one pointer for acquiring the boundary points visited around an obstacle

Procedure Traverse Boundary (current-coordinates)
Begin
Initial-coordinate = current-coordinate;
Boundary-coordinates:= Null;
Repeat
Move-to (current-coordinate) and mark the path of traversal;
Boundary-coordinates: = Boundary-coordinates $\cup$ {current-coordinate};
For (all possible unmarked set of point P)
Select the next point p ε P, such that
The perpendicular distance from the next point p to
Obstacle boundary is minimum;
Endfor
current-coordinate := next-coordinate;

Until current-coordinate = initial-coordinate
Return Boundary-coordinates;
End.

The above algorithm is self-explanatory and thus needs no elaboration. An algorithm for map building is presented below, where the procedure `Traversal Boundary` has been utilized.

Procedure Map Building (current-coordinate)
Begin
Move-to(current-coordinate);
Check-north-direction( );
If (new obstacle found) **Then do**
Begin
Current-obstacle = Traverse-boundary(new-obstacle-coordinate);
Add-obstacle-list (current-obstacle); //adds current obstacle to list//
Current-position = find-best-point (current-obstacle) // finding the best
        take off   point from the current obstacle//
Call Map-building (current-position);
End
Else do Begin
Check-north-east-direction ( );
If (new obstacle found) **Then do**
Begin
Current-obstacle = Traverse-boundary(new-obstacle-coordinate);
Add-obstacle-list (current-obstacle);
Current-position = find-best-point (current-obstacle);
Call Map-building (current-position);
End;
Else do Begin
Check east direction( );
//Likewise in all remaining directions//
End
Else backtrack to the last takeoff point on the obstacle (or the starting point);
End.

Procedure `Map Building` is a recursive algorithm that moves from an obstacle to the next following the depth-first traversal criteria. The order of preference of visiting the next obstacle comes from the prioritization of the directional movements in a given order. The algorithm terminates by back-tracking from the last obstacle to the previous one and finally to the start-ing point.

2.2.2 Explanation of the Algorithm

The procedure Map Building gradually builds up a tree, the nodes of which denote the obstacles/boundary visited. Each node in the tree keeps a record of the boundary pixel coordinates. The procedure expands the node satisfying the well-known depth-first strategy. For instance, let n_i be the current root node. The algorithm first checks whether there exists any obstacle in the north direction. In case it exists, the algorithm allows the robot to move to an obstacle to the north of n_i, say at point n_j. The boundary of the obstacle is next visited, by the robot, until it reaches the point n_j. The algorithm then explores the possibility of another obstacle to the north of n_j, and continues so until no obstacle is found to the north of a point, say n_k on an obstacle. Under this circumstance only, the algo-rithm checks for possible obstacle to the north-east of point n_k. It is thus clear how depth-first search has been incorporated in the proposed algo-rithm. It needs to be pointed out that once the robot visits the boundary of an obstacle, the corresponding pixel-wise boundary descriptors are saved in an array.

The algorithm terminates when it reaches an obstacle at a point n_i and moves around its boundary but couldn't trace any obstacle in any of the possible eight direction around n_i. It then backtracks to the nodes, from where it visited node n_i. Let n_j be that parent node of n_i in the tree. Again if there exist no obstacles in any of the possible eight directions around n_j, then it backtracks to the parent of n_j. The process of backtracking thus con-tinues until it returns to the starting point. The following properties envis-age that the proposed algorithm is complete and sound.

Property 1: The procedure Map Building is complete.

Proof: By completeness of the algorithm, we mean that it will visit all ob-stacles and the floor boundary before termination.

Let us prove the property by the method of contradiction, i.e. there remains one or more obstacles before the termination of the algorithm. Now, since the algorithm has been terminated, there must be several backtrackings from n_m to n_k, n_k to n_j, n_j to n_i and so until the root node of the entire tree is reached. This can only happen if no points on an unvisited obstacle is along any of the eight possible directions of the points on all visited obstacles. The last statement further implies that the unvisited obstacle, say O_X, is completely surrounded by other unvisited obstacles O_K such that $\cup\ O_k$ covers the entirety of O_K. Now, this too can only happen, if each of O_K is surrounded fully by other obstacles O_j. If the process continues this way, then the entire floor space will ultimately be covered by all unvisited obstacles, which however is a contradiction over the initial premise. So the initial premise is wrong, and hence the property follows.

Property 2: The procedure `Map Building` is sound.

Proof: The algorithm never generates a node corresponding to a point in a free space, as the algorithm visits the boundaries of the obstacles only, but not any free space. Hence the proof is obvious.

2.2.3 An Illustration of Procedure `Traverse Boundary`

This example illustrates how a robot moves around an obstacle by utilizing the procedure `Traverse Boundary`. Consider the rectangular obstacle and robot (encircled R) on the north side of the obstacle (shown in Fig. 2.4).

Let the robot's initial coordinates be $x = 75$, $y = 48$ (measured in 2D screen pixels). The sensor information from the robot's location in all directions is stored in an array as

N	NE	E	SE	S	SW	W	NW
(75,43)	(80,43)	(80,48)	(80,50)	(75,50)	(70,50)	(70,48)	(70,43)

The shaded portions are obstacle regions. So we choose the location (70, 48) (in the west direction) as the next location to move, since it is an obstacle-free point and it is next to an obstacle region. So the robot will move to a step ahead in the west direction.

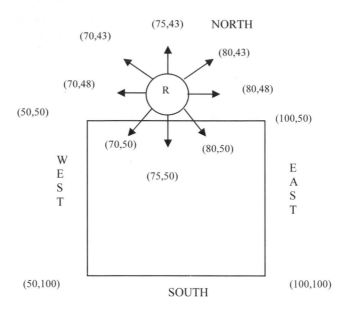

Fig. 2.4. The robot near a rectangular obstacle showing the sensory information in the eight specified directions

Let us consider the robot at another position (48, 48) shown in Fig. 2.5. The sensory information of the robot at this new location is given as follows.

N	NE	E	SE	S	SW	W	NW
(48,43)	(53,43)	(53,48)	(50,50)	(48,53)	(43,53)	(43,48)	(43,43)

From the above table, it is clear that the location (48, 53) is in the south direction, which is obstacle-free and it is next to the obstacle location (53, 53). So the next point to move is (48, 53) in the south direction. Likewise we repeat this process until we reach the initial coordinates (75, 48). The coordinates of the boundary of the obstacle are stored in a linked list, which is maintained in a general structure. Now the next task is to build the total map with the help of this boundary traversing algorithm, which is illustrated below.

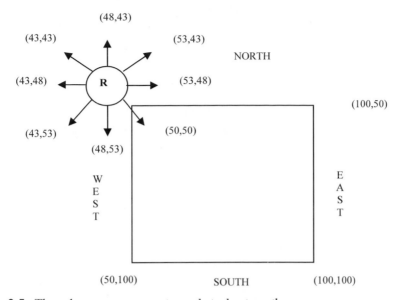

Fig. 2.5. The robot near a square-type obstacle at another location and the sensor information in the simulation

2.2.4 An Illustration of Procedure `Map Building`

An example is given here for creation of a linked list to record the visited obstacles and their boundaries of an environment shown in Fig. 2.6.

The searching process is started from the north direction of robot. If any obstacle is found, the robot will move to that obstacle and record the boundary coordinate information of that obstacle. Again it will start searching in the north direction from the recently visited obstacle. In this way it will go as deep as possible in the north direction only. If no new obstacle is available in the north direction, the robot will look for other directions, in order, for new obstacles. If any new one is found, the robot will visit it and move as deep as possible in the newly found direction. If in any case it cannot find any new obstacle, it will back-track to its parent obstacle and start looking in other directions, as shown in Fig. 2.7.

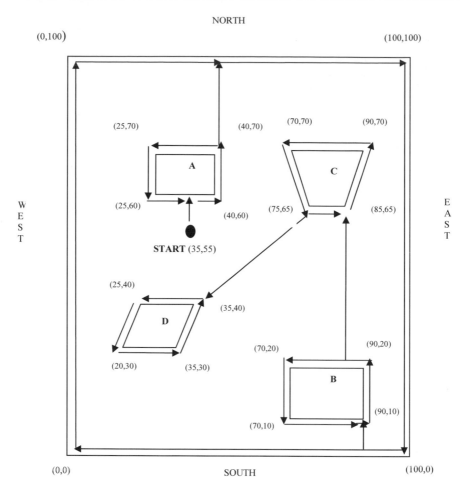

NORTH

(0,100)   (100,100)

(25,70)   (40,70)   (70,70)   (90,70)

A

C

W   (25,60)   (75,65)   (85,65)   E
E   (40,60)   A
S   S
T   T

START (35,55)

(25,40)

(35,40)

D

(20,30)   (35,30)   (70,20)   (90,20)

B

(90,10)

(70,10)

(0,0)   SOUTH   (100,0)

Fig. 2.6. The path traveled by the robot while building the 2D world map using depth-first traversal. The obstacles are represented with literals and the coordinates of the obstacles and the workspace is shown inside the braces

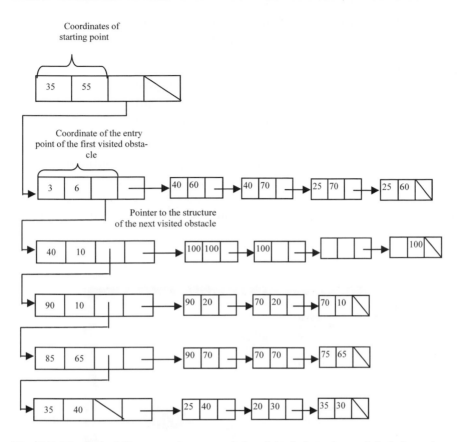

Fig. 2.7. The linked list created to record the visited obstacles and their boundary. Here all the points on the boundary visited by the robot have not been shown in the linked list in order to maintain clarity

2.2.5 Robot Simulation

The algorithm for map building has been simulated and tested by a C++ program. An artificial workspace has been created with nine obstacles along with a closed room, which is shown in Fig. 2.8. The workspace dimension is fixed by four corner points having coordinates (80, 80), (400, 80), (400, 400) and (80, 400) in a (640, 480) resolution screen. The dimensions of the obstacles, described by their peripheral vertices, are as follows:
Obstacle 1: (140,120), (170,100), (185,120), (175,140)
Obstacle 2: (240,120), (270,140), (225,164), (210, 135)

Obstacle 3: (178,160), (280,180), (185,200), (170,180)
Obstacle 4: (245,175), (285,200), (258,204), (230,190)
Obstacle 5: (310,215), (360,240), (330,270), (298,250)
Obstacle 6: (110,245), (130,225), (180,240), (130,280)
Obstacle 7: (230,258), (270,250), (250,280), (220,280)
Obstacle 8: (220,320), (230,300), (250,330), (230,340)
Obstacle 9: (190,330), (210,350), (180,370), (170,350)

The source code is available in Listing 2.1 at the website of the book. The dimension of the soft mobile object is 10 pixels in diameter. The soft object starts at position (100, 380), and moves as per the map building algorithm, which is displayed in Fig. 2.9 and the simulation results are shown in Fig. 2.10.

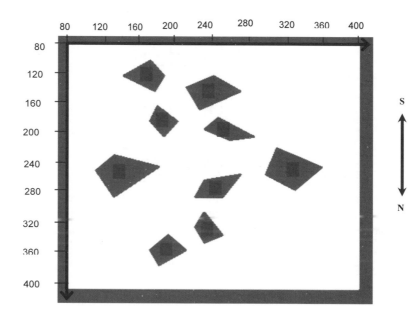

Fig. 2.8. A closed room workspace with nine convex obstacles

2.3 Execution of the Map Building Program

Following are the instructions to be followed for running this program.

```
Enter the starting X_position of Robot (80-400): 100
(enter)
Enter the starting Y_position of Robot (80-400): 380
(enter)
```

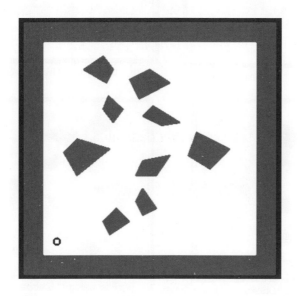

Fig. 2.9. The workspace along with the robot position

The simulation results given in Fig. 2.10 and the coordinates of the boundaries visited by the robot are given subsequently.

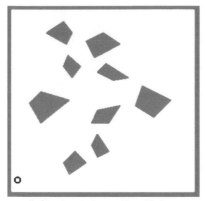

(i): Showing starting position of Robot

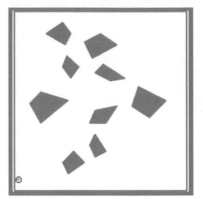

(ii): After visiting First Obstacle.

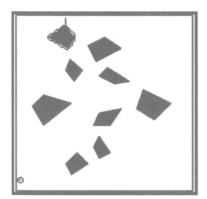

(iii): After visiting 2nd Obstacle

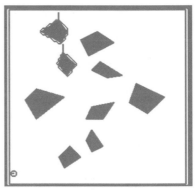

(iv): After visiting 3rd Obstacle

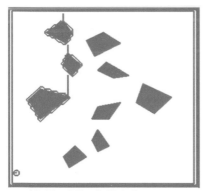

(v): After visiting 4th Obstacle

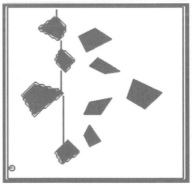

(vi): After visiting 5th Obstacle

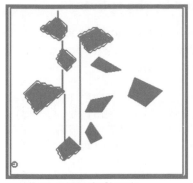

(vii): After visiting 6th Obstacle

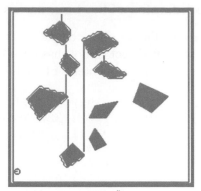

(viii): After visiting 7th Obstacle

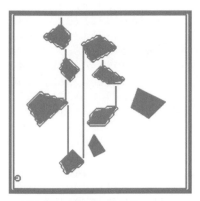

(ix): After visiting 8th Obstacle

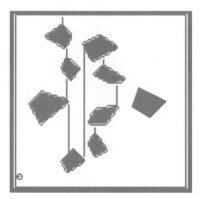

(x): After visiting 9th Obstacle

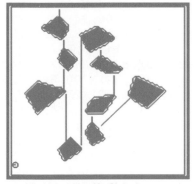

(xi): After visiting 10th Obstacle

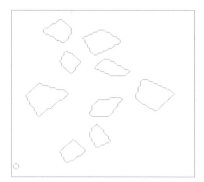

(xii): Silhouette of the workspace

Fig. 2.10. Experimental results in a closed workspace containing nine convex obstacles

Coordinates of the boundaries visited by the robot:

The coordinates of the boundaries of the obstacles and the room traversed by the soft object (robot) has been recorded in a file c:\coord.dat by the program. The contents of the file are given below.

```
(100,398) (104,398) (108,398) (112,398) (116,398) (120,398) (
124,398) (128,398) (132,398) (136,398) (140,398) (144,398) (1
48,398) (152,398) (156,398) (160,398) (164,398) (168,398) (17
2,398) (176,398) (180,398) (184,398) (188,398) (192,398) (196
,398) (200,398) (204,398) (208,398) (212,398) (216,398) (220,
398) (224,398) (228,398) (232,398) (236,398) (240,398) (244,3
98) (248,398) (252,398) (256,398) (260,398) (264,398) (268,39
8) (272,398) (276,398) (280,398) (284,398) (288,398) (292,398
) (296,398) (300,398) (304,398) (308,398) (312,398) (316,398)
(320,398) (324,398) (328,398) (332,398) (336,398) (340,398) (
344,398) (348,398) (352,398) (356,398) (360,398) (364,398) (3
68,398) (372,398) (376,398) (380,398) (384,398) (388,398) (39
2,398) (396,398) (400,398) (400,394) (400,390) (400,386) (400
,382) (400,378) (400,374) (400,370) (400,366) (400,362) (400,
358) (400,354) (400,350) (400,346) (400,342) (400,338) (400,3
34) (400,330) (400,326) (400,322) (400,318) (400,314) (400,31
0) (400,306) (400,302) (400,298) (400,294) (400,290) (400,286
) (400,282) (400,278) (400,274) (400,270) (400,266) (400,262)
(400,258) (400,254) (400,250) (400,246) (400,242) (400,238) (
400,234) (400,230) (400,226) (400,222) (400,218) (400,214) (4
00,210) (400,206) (400,202) (400,198) (400,194) (400,190) (40
0,186) (400,182) (400,178) (400,174) (400,170) (400,166) (400
,162) (400,158) (400,154) (400,150) (400,146) (400,142) (400,
138) (400,134) (400,130) (400,126) (400,122) (400,118) (400,1
14) (400,110) (400,106) (400,102) (400,98) (400,94) (400,90) (
400,86) (400,82) (396,82) (392,82) (388,82) (384,82) (380,82)
(376,82) (372,82) (368,82) (364,82) (360,82) (356,82) (352,82
) (348,82) (344,82) (340,82) (336,82) (332,82) (328,82) (324,8
2) (320,82) (316,82) (312,82) (308,82) (304,82) (300,82) (296,
82) (292,82) (288,82) (284,82) (280,82) (276,82) (272,82) (268
,82) (264,82) (260,82) (256,82) (252,82) (248,82) (244,82) (24
0,82) (236,82) (232,82) (228,82) (224,82) (220,82) (216,82) (2
12,82) (208,82) (204,82) (200,82) (196,82) (192,82) (188,82) (
184,82) (180,82) (176,82) (172,82) (168,82) (164,82) (160,82)
(156,82) (152,82) (148,82) (144,82) (140,82) (136,82) (132,82
) (128,82) (124,82) (120,82) (116,82) (112,82) (108,82) (104,8
2) (100,82) (96,82) (92,82) (88,82) (84,82) (80,82) (80,86) (80
,90) (80,94) (80,98) (80,102) (80,106) (80,110) (80,114) (80,1
18) (80,122) (80,126) (80,130) (80,134) (80,138) (80,142) (80,
146) (80,150) (80,154) (80,158) (80,162) (80,166) (80,170) (80
```

,174) (80,178) (80,182) (80,186) (80,190) (80,194) (80,198) (8
0,202) (80,206) (80,210) (80,214) (80,218) (80,222) (80,226) (
80,230) (80,234) (80,238) (80,242) (80,246) (80,250) (80,254)
(80,258) (80,262) (80,266) (80,270) (80,274) (80,278) (80,282
) (80,286) (80,290) (80,294) (80,298) (80,302) (80,306) (80,31
0) (80,314) (80,318) (80,322) (80,326) (80,330) (80,334) (80,3
38) (80,342) (80,346) (80,350) (80,354) (80,358) (80,362) (80,
366) (80,370) (80,374) (80,378) (80,382) (80,386) (80,390) (80
,394) (80,398) (84,398) (88,398) (92,398) (96,398) (100,398)
Obstacle Boundary
(168,99) (172,99) (176,103) (180,107) (180,111) (184,115) (18
8,119) (188,123) (184,127) (184,131) (180,135) (180,139) (176
,143) (172,143) (168,139) (164,139) (160,135) (156,131) (152,
131) (148,127) (144,127) (140,123) (136,119) (140,115) (144,1
15) (148,111) (152,107) (156,107) (160,103) (164,99) (168,99)
Obstacle Boundary
(176,161) (180,157) (184,161) (188,165) (192,169) (196,173) (
200,177) (204,181) (200,185) (196,189) (192,193) (192,197) (1
88,201) (184,201) (180,197) (176,193) (172,189) (172,185) (16
8,181) (168,177) (168,173) (172,169) (172,165) (176,161)
Obstacle Boundary
(180,238) (180,242) (176,246) (172,250) (168,254) (164,258) (
160,258) (156,262) (152,266) (148,270) (144,274) (140,274) (1
36,278) (132,282) (128,282) (124,278) (124,274) (120,270) (12
0,266) (116,262) (112,258) (112,254) (108,250) (108,246) (108
,242) (112,238) (116,234) (120,230) (124,226) (128,222) (132,
222) (136,222) (140,226) (144,226) (148,226) (152,230) (156,2
30) (160,230) (164,230) (168,234) (172,234) (176,234) (180,23
8)
Obstacle Boundary
(180,338) (184,334) (188,330) (192,330) (196,334) (200,338) (
204,342) (208,346) (212,350) (208,354) (204,358) (200,362) (1
96,362) (192,366) (188,370) (184,370) (180,374) (176,370) (17
6,366) (172,362) (172,358) (168,354) (168,350) (172,346) (176
,342) (180,338)
Obstacle Boundary
(212,142) (208,138) (208,134) (212,130) (216,130) (220,126) (
224,126) (228,122) (232,122) (236,118) (240,118) (244,118) (2
48,122) (252,126) (256,126) (260,130) (264,134) (268,134) (27
2,138) (272,142) (268,146) (264,146) (260,150) (256,150) (252
,154) (248,154) (244,158) (240,158) (236,162) (232,162) (228,
166) (224,166) (220,162) (220,158) (216,154) (216,150) (212,1
46) (212,142)
Obstacle Boundary
(244,174) (248,174) (252,174) (256,178) (260,182) (264,182) (
268,186) (272,190) (276,190) (280,194) (284,194) (288,198) (2
84,202) (280,206) (276,206) (272,206) (268,206) (264,206) (26

0,206) (256,206) (252,206) (248,202) (244,202) (240,198) (236
,198) (232,194) (228,190) (232,186) (236,182) (240,178) (244,
174)
Obstacle Boundary
(268,248) (272,252) (268,256) (268,260) (264,264) (260,268) (
260,272) (256,276) (252,280) (248,284) (244,284) (240,284) (2
36,284) (232,284) (228,284) (224,284) (220,284) (216,280) (22
0,276) (220,272) (220,268) (224,264) (224,260) (228,256) (232
,256) (236,252) (240,252) (244,252) (248,252) (252,252) (256,
248) (260,248) (264,248) (268,248)
Obstacle Boundary
(232,300) (236,304) (240,308) (240,312) (244,316) (248,320) (
248,324) (252,328) (252,332) (248,336) (244,336) (240,340) (2
36,340) (232,344) (228,340) (224,336) (224,332) (220,328) (22
0,324) (216,320) (220,316) (220,312) (224,308) (224,304) (228
,300) (232,300)
Obstacle Boundary
(297,251) (297,247) (297,243) (297,239) (301,235) (301,231) (
301,227) (305,223) (305,219) (305,215) (309,211) (313,215) (3
17,215) (321,219) (325,219) (329,223) (333,223) (337,227) (34
1,227) (345,231) (349,231) (353,235) (357,235) (361,239) (361
,243) (357,247) (353,251) (349,255) (345,259) (341,263) (337,
267) (333,271) (329,271) (325,271) (321,267) (317,267) (313,2
63) (309,259) (305,259) (301,255) (297,251)

2.4 Summary

The chapter presents a tool for the representation of the 2D environment of a mobile robot along with a simulation employing a depth-first search strategy.

3 Path Planning

3.1 Introduction

Path planning of mobile robots means to generate an optimal path from a starting position to a goal position within its environment. Depending on its nature, it is classified into offline and online planning. Offline planning determines the trajectories when the obstacles are stationary and may be classified again into various types such as obstacle avoidance; path traversal optimization; time traversal optimization. The obstacle avoidance problems deal with identification of obstacle free trajectories between the starting point and goal point. The path traversal optimization problem is concerned with identification of the paths having the shortest distance between the starting and the goal point. The time traversal optimization problem deals with searching a path between the starting and the goal point that requires minimum time for traversal. Another variant which handles path planning and navigation simultaneously in an environment accommodating dynamic obstacles is referred to as online navigational planning, which we will cover in the next chapter. Depending on the type of planning, the robot's environment is represented by a tree, graph, partitioned blocks, etc. Let us discuss the path planning problem, with suitable structures and representations.

3.2 Representation of the Robot's Environment

Let us first discuss the generalized Voronoi diagram (GVD) representation to find a path from a starting node to a goal node. The GVD describes the free space for the robot's movement in its environment. There exist various approaches to construct the GVD namely the potential field method [Rimon, 1992], two-dimensional cellular automata [Tzionas et al., 1997], and piecewise linear approximation [Takahashi, 1989].

3.2.1 GVD Using Cellular Automata

The construction process of GVD by cellular automata starts after representing the workspace as a rectangular grid. First the boundary grid cells of each obstacle and the inner space boundary of the environment is filled in, with a numeral say 1 (One). As the distance from the obstacles and the inner boundary increases, the coordinates of the space will be filled in, with gradually increasing numbers. The process of filling in the workspace by numerals is thus continued until the entire space is filled in. Next, the cells numbered with highest numerals are labeled and its neighborhood cells containing the same numerals or one less than that is labeled. The process of labeling is continued until each obstacle is surrounded by a closed chain of labeled numerals. Such closed curves are retained and the rest of the numerals in the space are deleted. The graph, thus constructed, is called the GVD. An example is shown in Fig. 3.1, which demonstrates the construction process of GVD.

Here the workspace is represented by 16 × 16 grid cells, with the obstacles in it. The cell distance from the obstacle as well as the boundary is calculated and denoted by that number. The collision-free path is traced by taking the maximum distance cell from the obstacle and the boundary. The robot will proceed in the shortest distance path, by employing a heuristic

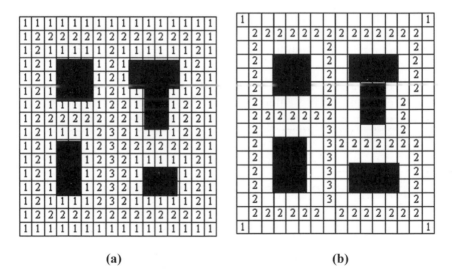

(a)   (b)

Fig. 3.1 (a) Cell distance calculation by cellular automata; **(b)** Collision free path

search technique. If it fails, it will reject that path and start navigating in the next shortest path, and so on.

3.3 Path Optimization by the Quadtree Approach

3.3.1 Introduction to the Quadtree

Region representation is an important aspect of image processing. In recent years, a considerable amount of interest has been developed in quadtrees. This is because of their hierarchical nature which causes a compact representation. It is quite efficient for a number of traditional image processing operations such as computing parameters, labeling connected components, finding the genuineness of an image, computing centroids and set properties.

Conventional path planning algorithms can be divided broadly into two categories. The first category makes trivial changes to the representation of the image map before planning a path. Although this method has a minimum cost of representation, it is rarely used for mobile robot navigation. One of the examples of this category is the cell decomposition approach. This approach consists of subdividing an environment into discrete cells of a predefined shape and size, such as a square, and then searching an undirected graph based on the adjacency relationships between the cells. This approach has the advantage of being able to generate accurate paths, although they are inefficient when environments contain large areas of obstacle-free regions. Its path planning cost increases with grid size, rather than with the number of obstacles present.

The second category makes an elaborate arrangement to convert to a representation which will be easier to analyze before planning a path. Free space methods, Voronoi methods, and medial axis transform methods are some examples. The practical shortcoming of such methods is that the cost of planning is still high, because of the representation conversion process involved.

The quadtree approach is a trade-off between these two categories. The hierarchical nature of the quadtree data structure makes it a popular choice for other applications because it is adaptive to the clutter of an environment. As the image map is converted into a smaller number of nodes, the quadtree gains a lot of computational saving. The following two aspects are considered for path planning purposes.

- A Mobile robot ordinarily negotiates any given path only once, which implies that it is more important to develop a negotiable path quickly than to develop an "optimal" path, which is usually an expensive affair.
- The Mobile robot should keep a safe distance from obstacles in the environment.

3.3.2 Definition

The quadtree is a tree, where each node has four child nodes. Any two-dimensional map can be represented in the form of quadtree by recursive decomposition [Davis, 1986]. Each node in the tree represents a square block of the given map. The size of the square block may be different from node to node. The nodes in the quadtree can be classified into three groups i.e. free nodes, obstacle nodes and mixed nodes.

- A *free node* is a node where no obstacles are present in the square region.
- An *obstacle node* is a node whose square region is totally filled with obstacles.
- A *mixed node's* square region is partially filled with obstacles.

3.3.3 Generation of the Quadtree

The generation process of the 2D map shown in Fig. 3.2(a) is first divided into four subsquare regions (four child nodes), namely NW, NE, SW, SE according to the directions. Here square regions NW, SW are fully occupied with obstacles (gray regions) and are called the "obstacle node", node NE does not have any obstacle in it and is called a "free node". The node SE is partially filled with the obstacle and is called "mixed node". The decomposition is shown in Fig. 3.2(b). Free nodes and obstacle nodes are not decomposed further and remain as leaf nodes. But the mixed node is subdivided into four subquadrants, which form children of that node. The decomposition procedure is repeated until either of the conditions mentioned below is satisfied.

1. The node is either a free node or an obstacle node.
2. The size of the square region represented by the child nodes is less than or equal to the size of the mobile robot.

Let us consider the map shown in Fig. 3.3, for the generation of the quad-tree. The data structure needed for a node is represented as given below in programming language C syntax.

```
Struct   node
 { node* pointer_to_child1;
   node* pointer_to_child2;
  node* pointer_to_child3;
  node* pointer_to_child4;
  node* pointer_to_parent_node;
  int        node_status;
  };
```

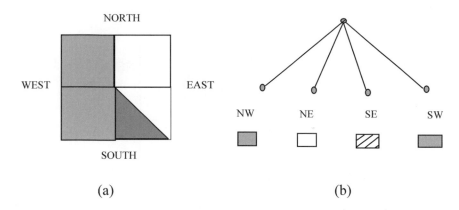

(a)   (b)

Fig. 3.2 (a) Representation of a simple 2D world map, in which the gray region represents obstacles; **(b)** Decomposition of the 2D world map into quadtree nodes. The type of each node is represented by the small box, with different fill patterns: gray color for obstacle nodes, white color for free nodes, and hatched boxes for mixed nodes

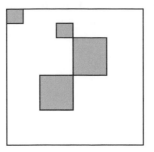

Fig. 3.3. A representative 2D world map of a robot, containing obstacles, denoted by the shaded regions

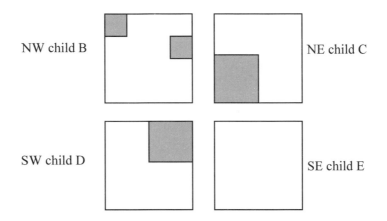

NW child B

NE child C

SW child D

SE child E

Fig. 3.4. The decomposition of the 2D world map

The gray square areas in Fig. 3.3 are regions occupied by obstacles. In the first stage of decomposition, the map is divided into four square regions of equal size as shown in Fig. 3.4. The root of the quadtree is the map itself and is denoted by A.

In the above decomposition the child E contains no obstacle and remains as a leaf node. The remaining nodes B, C, and D contain obstacles and are treated as mixed nodes. The quadtree after first decomposition is represented as in Fig. 3.5. The small square box under each node represents the status of the node, where the white box, gray box and the hashed line box represents a free node, obstacle node and a mixed node, respectively. The obstacle nodes are decomposed further, as shown in Fig. 3.6.

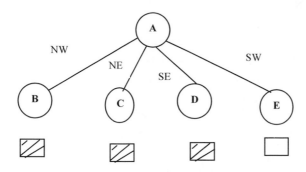

Fig. 3.5. The quadtree representation of the decomposition

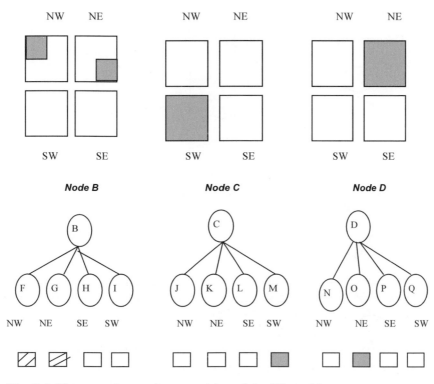

Fig. 3.6. The second stage decomposition of the 2D world map

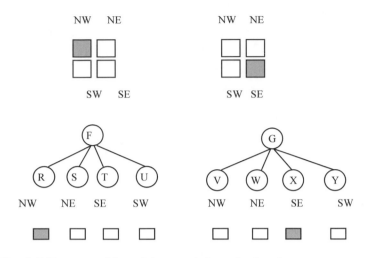

Fig. 3.7. Decomposition of the remaining mixed nodes

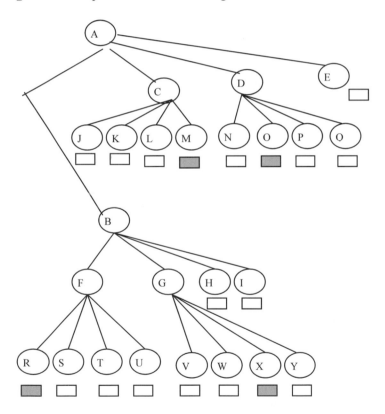

Fig. 3.8. The Qaudtree representation of the world map given in Fig. 3.3

After the second stage of decomposition, it is found that the nodes F and G are mixed nodes, which are decomposed further is shown in Fig. 3.7. After the third decomposition all nodes are either obstacle nodes or free nodes. This satisfies the first condition, mentioned earlier and terminates the generation process. The quadtree generated in this process is shown in Fig. 3.8.

After the generation of a complete quadtree, the robot generates alternative paths from one node to another leaf node by neighbor-finding algorithms [Samet, 1982] and then the optimum path is generated using the A* algorithm.

3.4 Neighbor-Finding Algorithms for the Quadtree

The two-dimensional image map is divided into a number of square blocks (may be of different sizes) while generating the quadtree. For path planning we need to move especially between adjacent blocks. Therefore, some technique is needed to find these adjacent blocks, called "neighbors" for a given square block. The original algorithm proposed by Hanan Samet [Samet, 1982], is discussed here in detail. The importance of these methods lies in their being a cornerstone of many of the quadtree algorithms. Since they are basically tree traversals with a visit at each node, they do not use the coordinate information, or the knowledge of the size of the image, which can be understood from the image shown in Fig. 3.9.

The neighbors of node D are regions B, E, F, and C in the north, east, south, and west directions, respectively. In our approach we do not take the corner neighborhood such as D and G, because of the possibility of the absence of a path between corner neighbors.

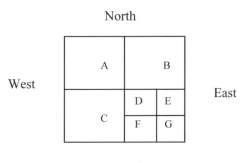

Fig. 3.9. A simple 2D world map

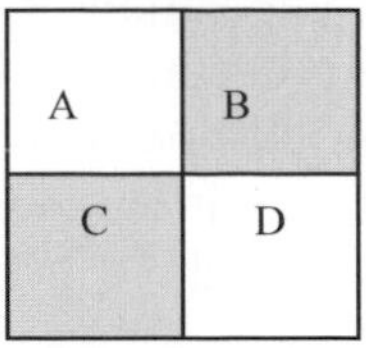

A	B
C	D

Fig. 3.10. A simple decomposed 2D world map

In Fig. 3.10, let us assume the robot is in the square region A, and the goal is to reach the square region D. If we take the corner neighborhood into account, the regions A and D will become neighbors. But there is no path to move into the region D from region A, since regions B and C are occupied with obstacles. This is the reason why the corner neighbors are neglected. As mentioned earlier, for a mixed node, one can get four immediate children in four directions. These are called NW, NE, SE, and SW, which are below in Fig. 3.11.

If P is a node, and I is a quadrant, then these fields are referenced as FATHER (P) and SON (P, I), respectively. One can determine the specific quadrant in which a node P lies relative to its father by the use of the function SONTYPE (P), which has the value of I, if

$$SON (FATHER(P),I) = P.$$

For instance assume Fig. 3.11 is a node named P, and the child node in the NW direction is named as Q. Then

FATHER (Q) = P.
SON(P,NW) =Q.
SONTYPE (P) = NW.

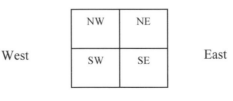

Fig. 3.11. Representation of children used in the algorithm

While generating the quadtree the "node status" is stored in every node. The integer values stored for the node status are

Node status = 0 = WHITE; If node is a free node.
Node status = 1 = BLACK; If node is an obstacle node.
Node status = 2 = GRAY;   If node is a mixed node.

The four boundaries of a node's square region are called N, E, S, and W for the north, east, south and west directions, respectively. The following predicates and functions are defined below, which will be used in the subsequent algorithm.

(i)   ADJ(B, I)   is true if and only if the quadrant I is adjacent to the boundary B of the node's block. For instance,

ADJ(W,NW) = TRUE and ADJ(W,NE) = FALSE.

(ii) REFLECT (B,I) yields the SONTYPE value of the block of equal size that is adjacent to the side B of a block having SONTYPE value I. For instance,

REFLECT(E,NW) = NE and REFLECT(N,SW) = NW.

REFLECT gives the mirror image of the node I in the direction B. For Fig. 3.9, the mirror image of child SW in the N (north) direction is NW and the mirror image of child SW in the E (east) direction is SE. These relations are represented in the tables shown in Fig. 3.12.

For the quadtree corresponding to a $2^n \times 2^n$ array, the root is at level n, and the node at level i, is at a distance n–i from the root of the tree. In other words, for a node at level i, we must ascend n–i FATHER links to reach the root of the tree. The following algorithm explains how to reach the neighboring node.

ADJ (S,Q)

		Quadrant 'Q'			
		NW	NE	SW	SE
	N	T	T	F	F
Side 'S'	E	F	T	F	T
	S	F	F	T	T
	W	T	F	T	F

REFLECT (S,Q)

		Quadrant 'Q'			
		NW	NE	SW	SE
	N	SW	SE	NW	NE
Side 'S'	E	NE	NW	SE	SW
	S	SW	SE	NW	NE
	W	NE	NW	SE	SW

Fig. 3.12. Predicate relations used in the neighbor finding algorithms

Algorithm   GTEQUAL_ADJ_NEIGHBOR (P,D)

Locate a neighbor of a node P in the horizontal or vertical direction D. If such a Node does not exist, then return NULL.

```
Begin
Value  node P;
Value direction D;
Node Q;
If ( not NULL(FATHER(P)) and  ADJ(D,SONTYPE(P))
Then
/* find common ancestor */
Q← GTEQUAL_ADJ_NEIGHBOR(FATHER(P),D)
Else
Q← FATHER(P);
```

```
/* follow the reflected path to locate the neighbor */
return (if (not NULL (Q) and node_status(q)=GRAY)
Then
SON(Q,REFLECT (D,SONTYPE(P)))
Else Q
End.
```

This algorithm will return a neighbor of greater or equal size. This is done by finding the common ancestor first. Next the path is retraced while making the mirror image move about an axis formed by the common boundary between the blocks associated with the nodes. The common ancestor is simple to determine. For instance, to find an eastern neighbor, the common ancestor is the first ancestor node which is reached via its NW or SW son. The procedure is shown in Fig. 3.13.

In Fig. 3.13, the eastern neighbor of the node A is G. It is located by ascending the tree until the common ancestor D, is found, from Fig. 3.14. This requires going through a NE link to B, a NE link to C, and a NW link to reach D. The node G is now reached by backtracking along the previous path with appropriate mirror image moves. This requires descending a NE link to reach E (since NW and NE are horizontal mirror images), a NW link to reach F and a NW link to reach G.

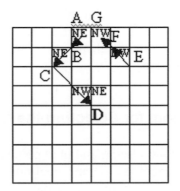

Fig. 3.13. Finding the neighbor of node A using a mirror image path from a common ancestor

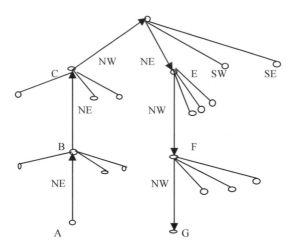

Fig. 3.14. Finding the neighbor of node A using a mirror image path from a common ancestor

After finding the equal-size neighbor, if it has any children, the smaller neighbors in it can be found by reaching every leaf node of it and, with the coordinate information available at that leaf node, checking the adjacency. The greater sized neighbor can be easily found from the given algorithm. It is the node which has no children, in the backtracking path, even though backtracking is not complete. After getting all neighbor nodes, the best neighbor node is selected among them, using an A* algorithm, which is explained in the next section.

3.5 The A* Algorithm for Selecting the Best Neighbor

From the knowledge of the starting and goal points, with a single traversal of the quadtree the stating node (S) and goal node (G) can be identified. Here the task is to find the minimum cost path between the starting node and goal node. For this purpose the A* algorithm is employed with the evaluation function f, of a node C which is defined as

$$f(c) = \ g(c) + h(c)$$

where $g(c)$ represents the cost of the path from S to C, and $h(c)$ represents the heuristic estimate of the cost of the remaining path from C to G. Since

the generated cost of a path should depend on both the actual distance traveled and the clearance of the path from the obstacles, g(c) is defined as

$$g(c) = g(p) + g'(p,c)$$

where g (p) is the cost of the path from S to C's predecessor P on the path
and g' (p, c) is the cost of the path segment between P and C.

The latter function, g'(p, c), in turn is defined as

$$g'(p, c) = D(p, c) + \alpha . d(c)$$

with D(c) representing the actual distance between nodes P and C, given
as half the sum of the node sizes, and d(c) representing the cost incurred by
including node C on the path. d(c) depends upon the clearance of the node
C from the nearby obstacles. A linear shape for the cost function d can be
chosen defining d(c) as

$$d(c) = O_{max} - O(c)$$

where O(c) is the distance of the node C from the nearest obstacle given by
the quadtree distance transform and O_{max} is the maximum such distance for
any node in the quadtree, so that d(c) is always positive. α in the equation
for g'(p, c) is a positive constant which determines how far the resultant
path will avoid obstacles.

The function h(c) is calculated as the Euclidean distance between mid-
points of the regions represented by C and G. Along with this single criterion, two more criteria can be included. They are the number of obstacles
intersecting the straight line path between C and G, and the second is the
total area of the obstacles intersecting the straight line path between C and
G. After calculating the evaluation function for all the neighbors of the
starting node, the lowest cost function is chosen among them and the node
(say X) corresponding to it is selected as the best node to move. The process of finding all neighbors and finding the best of them and then moving
to it, is repeated for the node X. This process is repeated until the goal
node is reached. The source code of the program Path Planning using
the quadtree method is available in Listing 3.1 at the website of the book
and the execution has been given in the next section.

3.6 Execution of the Quadtree-Based Path Planner Program

This program partitions a given workspace by the quadtree approach and determines the trajectory of the robot by employing a heuristic search in the tree. A sample run is given below, after executing the program.

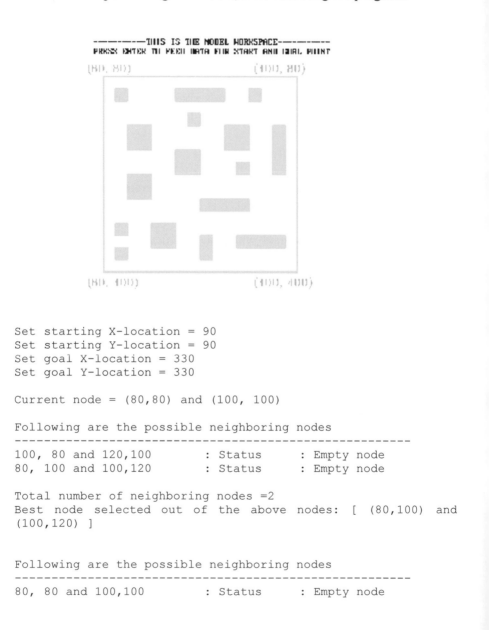

```
Set starting X-location = 90
Set starting Y-location = 90
Set goal X-location = 330
Set goal Y-location = 330

Current node = (80,80) and (100, 100)

Following are the possible neighboring nodes
---------------------------------------------------------
100, 80 and 120,100          : Status    : Empty node
80, 100 and 100,120          : Status    : Empty node

Total number of neighboring nodes =2
Best node selected out of the above nodes: [ (80,100) and
(100,120) ]


Following are the possible neighboring nodes
---------------------------------------------------------
80, 80 and 100,100           : Status    : Empty node
```

```
100, 100 and 120,120        : Status      : Occupied node
80, 120 and 120,160         : Status      : Empty node
```

Total number of neighboring nodes =3
Best node selected out of the above nodes: [(80,120) and (120,160)]

Following are the possible neighboring nodes

```
120, 120 and 160,160        : Status      : Empty node
80, 160 and 120,200         : Status      : Empty node
80, 100 and 100,120         : Status      : Empty node
100, 100 and 120,120        : Status      : Empty node
```

Total number of neighboring nodes =4
Best node selected out of the above nodes: [(120,120) and (160,160)]

Following are the possible neighboring nodes

```
120, 80 and 160,120         : Status      : Empty node
160, 120 and 200,160        : Status      : Empty node
120, 160 and 160, 200       : Status      : Occupied node
80, 120 and 120,160         : Status      : Empty node
```

Total number of neighboring nodes =4
Best node selected out of the above nodes: [(160,120) and (200,160)]

Following are the possible neighboring nodes

```
160, 80 and 200,120         : Status      : Empty node
160, 160 and 200,200        : Status      : Empty node
120, 120 and 160,160        : Status      : Empty node
200, 120 and 220,140        : Status      : Empty node
200, 140 and 220,160        : Status      : Empty node
```

Total number of neighboring nodes =5
Best node selected out of the above nodes: [(200, 140) and (220,160)]

Following are the possible neighboring nodes

```
200, 120 and 220,140        : Status      : Empty node
220, 140 and 240,160        : Status      : Occupied node
200, 160 and 240,200        : Status      : Empty node
160, 120 and 200,160        : Status      : Empty node
```

Total number of neighboring nodes =4
Best node selected out of the above nodes: [(200, 160) and (240,200)]

Following are the possible neighboring nodes

240, 160 and 280,200   : Status   : Empty node
200, 200 and 240,240   : Status   : Occupied node
160, 160 and 200,200   : Status   : Empty node
200, 140 and 220,160   : Status   : Empty node
220, 140 and 240,160   : Status   : Empty node

Total number of neighboring nodes =5
Best node selected out of the above nodes: [(200, 140) and
(220,160)]

Following are the possible neighboring nodes

240, 120 and 280,160   : Status   : Empty node
280, 160 and 320,200   : Status   : Occupied node
240, 200 and 280,240   : Status   : Empty node
200, 160 and 240,200   : Status   : Empty node

Total number of neighboring nodes =4
Best node selected out of the above nodes: [(240, 200) and
(280,240)]

Following are the possible neighboring nodes

240, 160 and 280,200   : Status   : Empty node
240, 240 and 280,280   : Status   : Empty node
200, 200 and 240,240   : Status   : Occupied node
280, 200 and 300,220   : Status   : Empty node
280, 220 and 300,240   : Status   : Empty node

Total number of neighboring nodes =5
Best node selected out of the above nodes: [(240, 240) and
(280,280)]

Following are the possible neighboring nodes

240, 200 and 280,240   : Status   : Empty node
280, 240 and 320,280   : Status   : Empty node
160, 240 and 240,320   : Status   : Empty node
240, 280 and 260,300   : Status   : Occupied node
260, 280 and 280,300   : Status   : Occupied node

Total number of neighboring nodes =5
Best node selected out of the above nodes: [(280, 240) and
(320,280)]

Following are the possible neighboring nodes

320, 240 and 400,320   : Status   : Empty node
240, 240 and 280,280   : Status   : Empty node
280, 220 and 300,240   : Status   : Empty node

```
300, 220 and 320,240      : Status    : Occupied node
280, 280 and 300,300      : Status    : Occupied node
300, 280 and 320,300      : Status    : Occupied node
```

Total number of neighboring nodes =6
Best node selected out of the above nodes: [(320, 240) and
(400,320)]

Following are the possible neighboring nodes

```
320, 200 and 360,240      : Status    : Empty node
360, 220 and 380,240      : Status    : Occupied node
380, 220 and 400,240      : Status    : Empty node
320, 320 and 340,340      : Status    : Empty node
340, 320 and 360,340      : Status    : Empty node
360, 320 and 380,340      : Status    : Empty node
380, 320 and 400,340      : Status    : Empty node
280, 240 and 320,280      : Status    : Empty node
300, 280 and 320,300      : Status    : Occupied node
300, 300 and 320,320      : Status    : Empty node
```

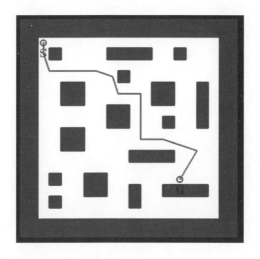

Total number of neighboring nodes =10
Best node selected out of the above nodes: [(320, 320) and
(340,340)]
Total time taken to search path = 966.1 msec.
Total path traversal = 476.42 units.

3.7 Summary

In quadtree-based path planning, the time of search is very smaller, as the number of nodes to be searched is considerably smaller. In fact, the number of leaf nodes in a quadtree of an image map having polygonal obstacles is approximately 2/3 $O(p)$ where p is the sum of the perimeters of the obstacles in terms of the lowest resolution units. The A* search will only have to deal with about $O(p)$ nodes in the case of a quadtree, instead of n^2 grid points in the case of a grid search method. Moreover a hierarchy of different levels of description of the space that is available with quadtrees enables us to search for a path close to the obstacles only when necessary. Corner clipping, inflexible paths are eliminated by considering only neighbors in horizontal and vertical directions. But the shortcoming of this method is that even though the generated path is an optimal one, it is with respect to the divided square blocks and not the exact path that a robot can travel.

4 Navigation Using a Genetic Algorithm

4.1 Introduction

When the map of the robot's environment is known, it can plan its trajectory before navigation to a predefined goal point from the starting point. But, many times the robot cannot decide about the entire trajectory before navigation, because the obstacles change their positions over time. Further, in a dynamic environment, which includes one or more mobile objects, it is useless to plan a path before navigation. In this situation, a robot plans a subpath and thus navigates to that point and the process continues until the robot reaches the destination. While moving towards the subgoal, if it discovers an obstacle, then it retraces back to the previous point and then replans for an alternative subpath. The process of replanning and navigation may continue until the robot reaches a given goal point [Patnaik et al., 1999c]. In this chapter navigational planning will be discussed using evolutionary algorithms [Patnaik et al., 1998].

The evolution program is a probabilistic algorithm, which maintains a population of individuals, $P(i) = \{ x_i^i, ..., x_n^i \}$ at iteration i. Each individual represents a potential solution to the problem at hand, and each solution x_i^t is evaluated to give some measure of its "fitness". Then, a new population at iteration (i + 1) is formed by selecting the better suited individuals. Some members of this population undergo transformations by means of unary transformations m_i (mutation), which create new individuals by a small change in a single individual ($m_i: S \rightarrow S$), and higher order transformations like c_j (crossover), which create new individuals by combining parts from several (two or more) individuals ($c_j: S \times S \rightarrow S$). After several generations, the program converges to a near optimum solution, hopefully representing the best individual. The structure of an evolution program is shown below.

Algorithm Evolution Program

```
Begin
i← 0;
Initialize population P_i;
Evaluate population P_i;
While (not termination-condition) do
For i=1 to n
i← i+1;
Select population P_i from previous population P_{i-1};
Apply genetic operators i.e. cross over and mutation on
population P_i;
Evaluation of population P_i by the predefined criteria;
End For;
End While
End;
```

Genetic algorithms have been successfully employed in various classical problems of AI such as intelligent search, optimization and machine learning. Let us first discuss genetic algorithms and their formulation in detail.

4.2 Genetic Algorithms

Genetic algorithms are inspired by Darwin's theory about evolution. They were invented by John Holland and developed by him and by his colleagues during 1975. According to Holland, the solution to a problem is evolved by genetic algorithms, rather than estimating it. The algorithm is started with a *set of solutions* (represented by *chromosomes*) called **a** *population*. Solutions from one population are taken and used to form a new population. This is done with the hope that the new population will be better than the old one. Solutions that are selected to form new solutions (*offspring*) are selected based on their fitness, i.e. the more suitable they are the more chances they have for reproduction. This process is repeated until some predefined condition is satisfied or the best solution is achieved.

The basic steps of a genetic algorithm are as follows.

Step 1: Generate a random population of n chromosomes out of the given problem. This is the most important step for the solution.
Step 2: Evaluate the fitness function $f(x)$ of each chromosome x in the population.
Step 3: Create the new population by repeating the following steps until the new population is complete:

(a) Select two parent chromosomes from a population according to their fitness. The better the fitness, the bigger the chance to be selected. This step is called **Selection**.

(b) **Crossover** the parents to form a new offspring or children, with a crossover probability. If no crossover was performed, the offspring are an exact copy of the parents.

(c) Mutate new offspring at each locus or position in the chromosome, with a mutation probability (this step is called **mutation**) and place these new offspring in the population.

(d) Use the newly generated population for a further run of the algorithm. This process is called **replacement**.

Step 4: If the predefined condition is satisfied then stop and return the best solution in current population or else **Goto Step 2.**

This is a generalized algorithm, but there are many variations, which can be implemented differently for different problems. Next the question is to create chromosomes, and what kind of encoding one should follow to select parents for crossover. This can be achieved in many ways, but the algorithm should select the better parents. While selecting the new population from the generated offspring, it may sometimes loose the best chromosome. In order to overcome this at least one best solution must be copied without changes to a new population, so that the best solution found at any iteration can survive to the end of the run.

4.2.1 Encoding of a Chromosome

The chromosome should in some way contain information about the solution which it represents. The most common way of encoding is a binary string, which can be represented as shown in Table 4.1.

Table 4.1. Encoding of a chromosome

Chromosome 1	1101100100110110
Chromosome 2	1101111000011110

Each chromosome has one binary string. Each bit in this string can represent some characteristic of the solution or the whole string can represent a number. Of course, there are many other schemes of encoding, which depend mainly on the problems to be solved. For instance, one can encode integer or real numbers directly, which is sometimes useful for some specific problems.

4.2.2 Crossover

Crossover selects genes from parent chromosomes and creates a new offspring. The simplest way to do this is to choose randomly some crossover point and copy everything before this point from first parent and then copy everything after the crossover point from the second parent. There exist also many complicated crossover like, multipoint and ring crossover, which depends on the encoding scheme of the chromosome. The specific crossover designed for a specific problem may not be suitable for other problems. A sample cross over operation is shown in Table 4.2.

Table 4.2. Crossover operation (| is the crossover point)

Chromosome 1	**01011** | **00100111110**
Chromosome 2	11011 | 11000011010
Offspring 1	**01011** | 11000011010
Offspring 2	11011 | **00100111110**

4.2.3 Mutation

After a crossover is performed, mutation is done, in order to prevent all solutions in the current population falling into a local optimum of the solved problem. Mutation changes a bit randomly in the new offspring; in other word it does the fine-tuning. For binary encoding a few randomly chosen bits may be changed from 1 to 0 or from 0 to 1. Mutation can then be followed as shown in Table 4.3. The mutation also depends on the encoding scheme as well as the crossover.

Table 4.3. Mutation scheme

Original offspring 1	01011 11000011010
Original offspring 2	11011 0 0100111110
Mutated offspring 1	01011 11001011010
Mutated offspring 2	11011 0 0100110110

4.2.4 Parameters of a GA

There are two basic parameters of a GA, i.e. the crossover probability and
the mutation probability. The *crossover probability* says how often crossover will be performed. If there is no crossover, i.e. 0% probability, the
offspring is an exact copy of the parents. If there is full crossover, or 100%
probability, then all offspring are made by crossover. The objective of the
crossover is that new chromosomes will have good parts of old chromosomes and some of the new chromosomes which are better will be
evolved. However, it is good to let some parts of the population survive to
the next generation, which are really better chromosomes. The *mutation
probability* says how often parts of the chromosome will be mutated. If
there is no mutation, or 0% probability, offspring are taken after crossover
or copied without any change. If the mutation probability is 100%, the
whole chromosome is changed. Mutation is made to prevent the GA falling into local extremum, but it should not occur very often, otherwise the
GA will behave as *random search*.

There are also some other parameters of a GA, one of which is *population size,* which specifies how many chromosomes should be in the population in one generation. If there are too few chromosomes, the GA have
few possibilities to perform a crossover and only a small part of the search
space is explored. On the other hand, if there are too many chromosomes,
the evolution process slows down.

4.2.5 Selection

Selection is the process by which chromosomes are selected from the
population for the crossover. The main problem is how to select these
chromosomes. According to Darwin's evolution theory the best ones
should survive and create new offspring. There are many schemes to select
the best chromosomes, for example roulette wheel selection, Boltzman selection, tournament selection, rank selection, steady state selection, and
some others. Some of them are covered here.

Elitism: While creating new population by crossover and mutation,
there is more chance that the best chromosome is lost. Elitism is the name
of a method that first copies the best chromosome or a few best chromosomes to the new population. The rest is done by the process of crossover
and mutation. Elitism can very rapidly increase the performance of the
GA, because it prevents losing the best found solution.

4.3 Navigation by a Genetic Algorithm

The current literature on robotics [Lin et al., 1994; Michalewicz, 1986; Trojanowski, 1997; Xiao, 1997] has established that genetic algorithm are useful tools for robot navigation. Michalewicz [Xiao, 1997] first successfully applied a GA [Goldberg, 1989] in mobile robot navigation. In their model, Michalewicz considered a set of operators including crossover and mutations. An operator is selected based on its probability of occurrence and the operation is executed. The fitness evaluation function is then measured and proportional selection is employed to get the population in the next generation. This model is ideal for a static environment, but in case of a dynamic environment much of the computation time will be wasted for planning a complete path, which later is likely to be disposed of. An alternative solution in this situation can be selection of a path segment from the sensory reading after each genetic evolution. This can be extremely fast and thus can take care of movable obstacles having speed less than or equal to that of the robot.

4.3.1 Formulation of Navigation

For the selection of the chromosome or to set up an initial population, the sensor information is taken into account and the coordinates obtained from these sensors are used to set up the initial population. This formulation ensures that all the initial populations are feasible, in the sense that they are obstacle-free points and the straight path segments between the starting point and via points are also obstacle free. Since the path segment to the next point is evaluated after each genetic evolution, the data structure to represent the chromosome becomes very simple, as shown in Fig. 4.1.

Here (X_i, Y_i) is the starting point and (X_j, Y_j) is one of the 2D points, obtained from the sensor information, where chromosomes form the initial population. Next crossover is done among the initial population. It has been observed that if the cross-site is chosen randomly, then most of the offspring generated out of the crossover are not feasible, as those paths

Fig. 4.1. Representation of the chromosome of a single path segment from the sensory readings

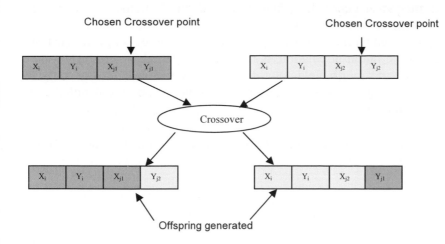

Fig. 4.2. The crossover operation used in the proposed algorithm

may either encounter obstacles or fall outside the workspace. Hence integer crossover is chosen instead of binary crossover. In Fig. 4.2 the crossover point is set between the third and the fourth field for each pair of chromosomes and the new population is generated. For the newly generated populations the feasibility is estimated, i.e. whether these paths are obstacle free or not. The next, mutation is performed for fine-tuning the path to avoid the sharp turns.

The estimation of fitness of each and every chromosome is done subsequently, out of the total population (both for the initial and new populations), which invloves finding the sum of the Euclidean distance from the starting point to the coordinate obtained from the sensor information and the distance from that point to the goal point.

Fitness of a chromosome $(X_i, Y_i, X_{jk}, Y_{jk}) =$

$$\frac{1}{(\text{distance between } (X_i \; Y_i) \; \& \; (X_{jk}, Y_{jk})) + (\text{distance between } (X_{jk}, Y_{jk}) \; \& \; (X_{goal}, Y_{goal}))}$$

$\forall k$, generated after the crossover.

The best-fit chromosome is evaluated, after finding the fitness value of each chromosome. The best-fit chromosome represents the predicted optimal path segment, towards the goal. A near optimal intermediate point is found after each generation. The third and fourth fields of the best-fit chromosome become the next intermediate point to move and the starting point is updated with the best-fit point. The whole process of the GA, from setting up the initial population, is repeated until the best-fit chromosome has its third and fourth field equal to the x- and y-coordinates of the goal location. The algorithm is formally presented below and the detailed code is given in the next section.

Procedure for Navigational Planning Using Genetic Algorithm

```
//  (xi, yi)  = starting point;
   (xg, yg) =goal point;    //
 add path-segment to path-list (xi,yi) ;
```

Repeat

i)   Initialization:
      Get sensor information in all possible directions
      (x_{j1}, y_{j1}), (x_{j2}, y_{j2}),….(x_{jn}, y_{jn}).
      Form chromosomes like (x_i, y_i, x_j, y_j);

ii)   Crossover:
      Select crossover point randomly on the third and
      the fourth fields of the chromosome.
      Allow crossover between all chromosomes and get
      new population. $(x_i, y_i, x_{j1}, y_{j1})$, $(x_i, y_i, x_{j2}, y_{j2})$,
      $(x_i, y_i, x_{j1}{}^i, y_{j1}{}^i)$, $(x_i, y_i, x_{j2}{}^{ii}, y_{j2}{}^{ii})$;

iii) Mutation:
      Select a mutation point in bitstream randomly and
      complement that bit position for every chromosome.

iv) Selection:
      Discard   all   chromosomes (x_i, y_i, x_j, y_j) from
      population   whole   line   segment   is   on   obstacle
      region
      For all chromosomes in population   find fittness
      using Fittness(x_i, y_i, x_j, y_j) = 1/ $((x_j-x_i)^2+(y_j-y_i)^2$
      $+ (x_g-x_j)^2+(y_g-y_j)^2$);
      Identify the best fit chromosome $(x_i, y_i, x_{bf}, y_{bf})$;
      Add to path-list(x_{bf}, y_{bf});
      $x_i = x_{bf}$; $y_i = y_{bf}$;
      End for,
      Until ($x_i = x_g$) && $(y_i = y_g)$;
End.

4.4 Execution of the GA-Based Navigation Program

Robot navigation by a genetic algorithm is simulated and tested by the C++ program. The source code is available in Listing 4.1 at the website of the book. An artificial workspace has been created with nine obstacles along with a closed workspace. The workspace dimension is fixed by four corner points having coordinates (80,80), (400,80), (400,400) and (80,400) in a (640,480) resolution screen. The dimensions of the obstacles, described by their peripheral vertices are as follows:

Obstacle A: (120,130), (180,130), (180,160), (120,160)
Obstacle B: (120,190), (140,220), (120,250), (100,220)
Obstacle C: (200,180), (250,180), (230,220), (180,220)
Obstacle D: (290,160), (320,160), (320,250), (290,250)
Obstacle E: (160,270), (230,270), (210,310), (180,310)
Obstacle F: (250,250), (290,250), (290,270), (250,270)
Obstacle G: (330,220), (360,220), (360,320), (330,320)
Obstacle H: (150,310), (220,380), (150,350)
Obstacle I: (260,330), (330,330), (330,380), (260,380)

The dimension of the soft mobile object is 10 pixels in diameter. Out of nine obstacles only obstacle F can change its position. The soft object starts at a position (130,370), and moves to a goal position (300,100) by the GA-based algorithm as shown in Fig. 4.3.

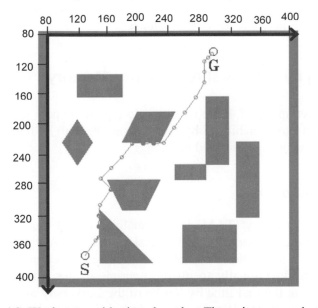

Fig. 4.3. Workspace with nine obstacles. The point-mass robot navigates from the starting point to the goal position smoothly

In Fig. 4.4, the obstacle F has changed to a new position [(250,180), (290,180), (290,200), (250,200)], which blocks the earlier path to the goal. In this situation, the soft mobile object navigates towards the goal by GA, after roving around the blocked path.

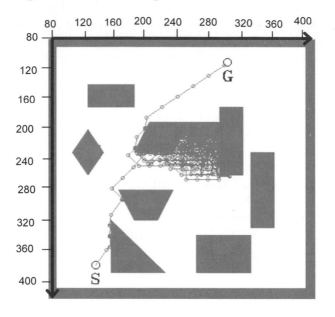

Fig. 4.4. Obstacle F has moved to a new location blocking the free path to the goal. The robot navigates in an alternative path using EA

4.5 Replanning by Temporal Associative Memory

In the situation shown in Fig. 4.4, the robot has to replan the path with the new situation, which can be achieved by memorizing the path by temporal associative memory which is described here.

4.5.1 Introduction to TAM

Temporal associative memory (TAM) is a specialized neural topology of *bi-directional associated memory* (BAM) [Kosko, 1987]. BAM [Kosko, 1988] is realized with a two-layer neural net, where each neuron at one layer is connected bi-directionally to all the neurons in the other layer. BAM was proposed for the first time by Kosko of the University of Southern California in the late 1980s [Kosko, 1987]. In his elementary model he considered a two-layered neural net, where neurons in each layer are connected to the neurons in the other layer through bi-directional links. The signal associated with the neurons can assume $\{-1, +1\}$ values and the weights, describing the connectivity between the neurons possess signed integer values. The basic problem in BAM was to design a single set of weight matrix W, such that the difference between the transformed vector and the output vector is a minimum.

$$\sum_{\forall i}(A_i W - B_i) + \sum_{\forall i}(B_i W^T - A_i) \tag{4.1}$$

Kosko considered a Lyapunov function, describing a nonlinear surface to show that there exist a single matrix W, for which the minima of expression (4.1) can be attained. This also proves that if B_i can be computed by taking the product of A_i and W, A_i too can be evaluated by multiplying B_i by W^T. The extension of the concept of BAM for memorizing a sequence of causal events is known as *temporal associative memory* (TAM) [Kosko, 1988]. For instance, given a set of weight matrix W, if one knows the causal relationships:

$A_1 \rightarrow A_2$
$A_2 \rightarrow A_3$
....
....
$A_{n-1} \rightarrow A_n$

then A_n can be inferred from A_1 through the chain sequence $A_1 \rightarrow A_2 \rightarrow A_3 \rightarrow .... \rightarrow A_n$. Now by remembering the weight matrix W and the input vector A_1, one can reconstruct the entire chain leading to A_n. Now, let us consider the case where there exists a bifurcation in the chain at the event A_j. The sequence describing the bifurcation is.

$A_1 \rightarrow A_2 \rightarrow ......\rightarrow A_j \rightarrow A_{j+1} \rightarrow .... \rightarrow A_{n+1}$

$\searrow B_1 \rightarrow B_2 \rightarrow B_3 \rightarrow .... \rightarrow B_k$

Let us assume that the whole sequence from A_i through A_n and from A_j through B_n is stored. Now it has been detected that the sequence A_{j+1} to A_n is invalid. The system under this configuration will backtrack from A_{j+1} to A_j by the operation $A_j = A_{j+1} .W^T$ and then proceed through an alternative sequence through B_i by the transformation $B_i = A_j.W'$. This property of TAM encourages employing it in replanning of the navigation of the mobile robot. Thus, it may be inferred that TAM may be useful for re-planning of mobile robot navigation in a dynamic world. Here the robot has to determine all possible paths between each pair of given locations, and encode it for subsequent usage. For instance, to traverse the path from A_1 to A_n, the robot has to memorize only the weight matrix W, based on which it can easily evaluate the entire trajectory passing through A_1 and A_n. In fact, we can derive A_2, A_3,......A_n by post multiplying A_1 by W, W^2, ...W^{n-1}. For determining the backtrack path from known A_n, we could go on multiplying by W^T. Because of the inherent feature of bi-directionality, the navigational replanning can be easily formulated with TAM.

4.5.2 Encoding and Decoding Process in a Temporal Memory

Once the possible path segments are generated by the EA navigator, the robot replanning can be encoded by using TAM. The encoding scheme in the present context is to evaluate the weight matrix W that satisfies the criterion of minimality of the function.

$$\sum_{\forall i}(A_i W - A_{i+1}) + \sum_{\forall i}(A_{i+1} W^T - A_i) \tag{4.2}$$

The decoding process means evaluating A_{i+1}, when A_i is known or vice-versa. Let us represent the motion of an autonomous vehicle on a path by a set of ordered vectors, such as $A_1 \rightarrow A_2 \rightarrow ......A_k \rightarrow A_{k+1} \rightarrow ......\rightarrow A_n$, assuming the temporal patterns are finite and discrete. The feature of the bi-directionality in BAM is being utilized here to memorize the associative matrix between $A_i \rightarrow A_{i+1}$. The following steps can be used to encode the following weight matrix of the TAM.

$$W = \sum_i^n X_i^T \cdot X_{i+1} \tag{4.3}$$

1. Binary vectors are converted to bipolar vectors
2. The contiguous relationship $A_i \rightarrow A_{i+1}$ is memorized by forming the correlation matrix $W_i = X_i^T \cdot X_{i+1}$
3. All the above correlation matrices are added point-wise to give

Decoding of the output vector A_{i+1} is estimated by vector multiplication of A with W and applying the following threshold function

$$A_{i+1}^{k} = \begin{cases} 1, & \text{if} \quad A_i W_i \geq 0 \\ 0, & \text{if} \quad A_i W_i < 0 \end{cases}$$

$$A_i^{k} = \begin{cases} 1, & \text{if} \quad A_{i+1} W_i^{T} > 0 \\ 0, & \text{if} \quad A_{i+1} W_i^{T} \leq 0 \end{cases}$$

4.5.3 An Example in a Semi-dynamic Environment

The trajectories traversed by the soft mobile object along two alternative paths, shown in Fig. 4.5 are represented by the following set of sequences in decimal numbers and later converted to binary values, where the number of bits depends on the length and resolution.

$A_1=(X_1,Y_1)=(1,1)=(0\ 0\ 1\ 0\ 0\ 1);\quad A_2=(X_2,Y_2)=(1,2)=(0\ 0\ 1\ 0\ 1\ 0);$
$A_3=(X_3,X_3)=(1,3)=(0\ 0\ 1\ 0\ 1\ 1);\quad A_4=(X_4,Y_4)=(2,4)=(0\ 1\ 0\ 1\ 0\ 0);$
$A_5=(X_5,Y_5)=(3,4)=(0\ 1\ 1\ 1\ 0\ 0);\quad A_6=(X_6,Y_6)=(4,4)=(1\ 0\ 0\ 1\ 0\ 0);$
$A_7=(X_7,Y_7)=(5,5)=(1\ 0\ 1\ 1\ 0\ 1);\quad A_8=(X_8,Y_8)=(5,6)=(1\ 0\ 1\ 1\ 1\ 0);$
$A_9=(X_9,Y_9)=(6,7)=(1\ 1\ 0\ 1\ 1\ 1);\quad A_{10}=(X_{10},Y_{10})=(7,7)=(1\ 1\ 1\ 1\ 1\ 1)$

Encoding the above sequences into bipolar values is necessary according to the TAM procedure, which is derived below:

$A_1=(X_1,Y_1)=(-1\ -1\ 1\ -1\ -1\ 1);\quad A_2=(X_2,Y_2)=(-1\ -1\ 1\ -1\ 1\ -1);$
$A_3=(X_3,X_3)=(-1\ -1\ 1\ -1\ 1\ 1);\quad A_4=(X_4,Y_4)=(-1\ 1\ -1\ 1\ -1\ -1);$
$A_5=(X_5,Y_5)=(-1\ 1\ 1\ 1\ -1\ -1);\quad A_6=(X_6,Y_6)=(1\ -1\ -1\ 1\ -1\ -1);$
$A_7=(X_7,Y_7)=(1\ -1\ 1\ 1\ -1\ 1);\quad A_8=(X_8,Y_8)=(1\ -1\ 1\ 1\ 1\ -1);$
$A_9=(X_9,Y_9)=(1\ 1\ -1\ 1\ 1\ 1);\quad A_{10}=(X_{10},Y_{10})=(1\ 1\ 1\ 1\ 1\ 1)$

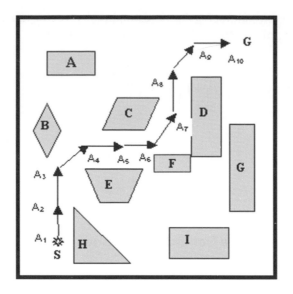

(a)

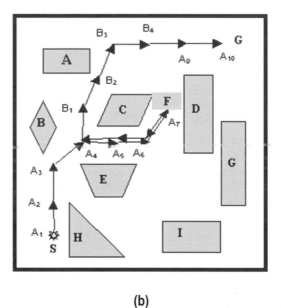

(b)

Fig. 4.5 (a) Sample path between starting position (S) and goal position (G) though a charted path A_1, A_2, ...A_{10}; **(b)** mobile object traverses in an alternate path after backtracking from node A_7 to A_4, as the semi-dynamic obstacle F has shifted its position and then navigates through the charted path from B_1 to B_4 and then takes the normal path to reach the goal within the workspace

The calculation of a sample correlation matrix W_1 is shown below.

$$W_1 = A_1^T . A_2 \quad = (\text{-}1 \text{ -}1 \text{ } 1 \text{ -}1 \text{ -}1 \text{ } 1)^T . (\text{-}1 \text{ -}1 \text{ } 1 \text{ -}1 \text{ } 1 \text{ -}1)$$

$$= \begin{bmatrix} 1 & 1 & -1 & 1 & -1 & 1 \\ 1 & 1 & -1 & 1 & -1 & 1 \\ -1 & -1 & 1 & -1 & 1 & -1 \\ 1 & 1 & -1 & 1 & -1 & 1 \\ 1 & 1 & -1 & 1 & -1 & 1 \\ -1 & -1 & 1 & -1 & 1 & -1 \end{bmatrix}$$

The weight matrix W is estimated by adding the contiguous matrices $W_1, W_2, \ldots\ldots W_8$.

$$W = W_1 + W_2 + W_3 + W_4 + W_5 + W_6 + W_7 + W_8 + W_9$$
$$= A_1^T . A_2 + A_2^T . A_3 + A_3^T . A_4 + A_4^T . A_5 + A_5^T . A_6 + A_6^T . A_7 + A_7^T . A_8 + A_8^T . A_9 + A_9^T . A_{10}$$

$$= \begin{bmatrix} 7 & 1 & 1 & 3 & 3 & 5 \\ 1 & 3 & -1 & 1 & -3 & -1 \\ -1 & -3 & -3 & -1 & 3 & -3 \\ 7 & 1 & 1 & 7 & -1 & 1 \\ 1 & 5 & -3 & -1 & 3 & 5 \\ -1 & 1 & 1 & -1 & 3 & -3 \end{bmatrix}$$

Each successive step of the movement can be estimated by the expression $A_i.W$, which can be easily verified. The second objective of the net is to backtrack, in the presence of an obstacle on a particular route. For instance, when the robot finds an obstacle after the seventh step of movement on the right-hand path of obstacle C, then it retraces back to node A_4, which is a junction and takes an alternative path by means of an alternative TAM weight matrix W', which has been estimated by considering the four nodes which surround the obstacle C, on the top and left side. Here from the node A_4, two alternate weight matrices are available. If the robot finds an obstacle in one of the paths, it will backtrack and proceed again from the node A_4 on the other path. The weight W' can be estimated as follows by considering the node points:

$A_4 = (2,4) = (0\ 1\ 1\ 1\ 0\ 0\ );$ $B_1 = (2,5) = (\ 0\ 1\ 0\ 1\ 0\ 1);$

$B_2 = (3,6) = (0\ 1\ 1\ 1\ 1\ 0);$ $B_3 = (4,7) = (1\ 0\ 0\ 1\ 1\ 1);$

$B_4 = (5,7) = (1\ 0\ 1\ 1\ 1\ 1);$ $A_9 = (6,7) = (1\ 1\ 0\ 1\ 1\ 1);$

$A_{10} = (7,7) = (1\ 1\ 1\ 1\ 1\ 1)$

Encoding the above sequences into bipolar values gives the following:

$A_4 = (-1\ 1\ 1\ 1\ -1\ -1\ );$ $B_1 = (\ -1\ 1\ -1\ 1\ -1\ 1);$

$B_2 = (-1\ 1\ 1\ 1\ 1\ -1);$ $B_3 = (1\ -1\ -1\ 1\ 1\ 1);$

$B_4 = (1\ -1\ 1\ 1\ 1\ 1);$ $A_9 = (1\ 1\ -1\ 1\ 1\ 1);$

$A_{10} = (1\ 1\ 1\ 1\ 1\ 1)$

$$W' = A_4^{\mathrm{T}}.\,B_1 + B_1^{\mathrm{T}}.\,B_2 + B_2^{\mathrm{T}}.\,B_3 + B_3^{\mathrm{T}}.\,B_4 + B_4^{\mathrm{T}}.\,A_9 + A_9^{\mathrm{T}}.\,A_{10}$$

$$= \begin{bmatrix} 4 & 0 & 2 & 0 & 2 & 2 \\ -2 & 2 & 0 & 2 & 0 & 0 \\ 0 & 0 & -6 & 0 & -2 & 2 \\ 2 & 2 & 0 & 6 & 4 & 4 \\ 6 & -2 & 0 & 2 & 4 & 4 \\ 2 & 2 & 4 & 2 & 4 & 0 \end{bmatrix}$$

4.5.4 Implications of Results

The previous simulation shown in Figs. 4.3 and 4.4 was repeated after memorizing path segments with the help of TAM. Fig. 4.6 shows that the robot is momentarily blocked by the obstacle F, and then finds a path by utilizing the TAM matrix for different path segments. The final route is shown in Fig. 4.7. While training the W matrix for different path segments, only eight successive node points are considered for a better approximation.

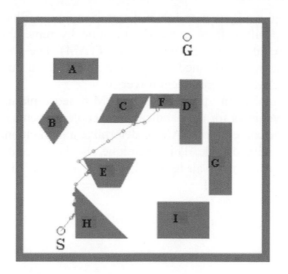

Fig. 4.6. Robot is blocked momentarily by obstacle F

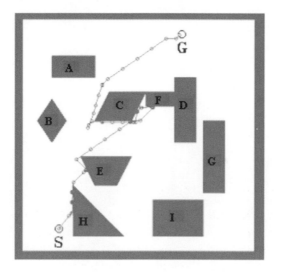

Fig. 4.7. Robot finds an alternate path by the TAM matrix to reach the goal

4.6 Summary

The chapter covers the interesting issue of replanning and navigation of a mobile robot, in the case of an environment filled with semi-dynamic obstacles. The robot first explores the environment by using a genetic algorithm-based navigation and then determines the possible trajectories to reach a goal position from a starting position by remembering a simple matrix using temporal associative memory (TAM). The possible path segments between two predefined positions are memorized through few node points using TAM, for subsequent traversal.

The most significant application of TAM-based replanning is in automatic guided vehicles for the handicapped people. Given a set of nodes n_1, n_2, ...n_n, one can employ TAM to determine a path emanating from one node to another fixed node. This technique also has potential applications in (i) handling cargo in an air terminal, and (ii) medicine distribution to patients in an indoor hospital.

5 Robot Programming Packages

5.1 Introduction

Essentially a robot is an assembly of mechanical actuators and sensors which are driven by their respective control circuitry, for example, the driving circuitry of a servo motor, the amplifier circuits of a sonar device, the electronic receptors of a camera, etc. Ultimately an entire program drives this hardware. As a programmer, however, one would not be as interested in wasting time and energy in dealing with the intricacies of the low-level functioning of the robot, as one should be in trying to manipulate the behavior of the robot at a higher level, guided by the task. Experimental robots usually have microcontroller driven hardware circuitry and the microcontroller takes care of the low-level device control. Besides that, it needs to act as the interface between the robot actuators and sensors and the user programs. The operating system of the robot serves this purpose, by interpreting the programs running on the microcomputer of the robot and mapping them into a sequence of instructions for the microcontroller. Similarly the feedback by the microcontroller is transferred to the user programs.

Here the programmer again faces the problem of knowing and repeatedly using code or commands for interacting with the operating system of the microcontroller. To make the task simpler for the programmer, the basic functionality is wrapped into classes or functions and is made available to the programmer in the form of new programming languages or libraries for existing languages such as C++ or JAVA. Some examples are cited here such as the ARIA library (for C++), the Saphira and Colbert programming environments to handle motor and sensor functions, the SVS library (for C and C++) to handle camera functions, the BotSpeak library (for C) for speech recognition and synthesis. As C++ is a popular and powerful OOP language, the programs discussed in this book have been tested in C++. The ARIA library for C++ has been used to implement modules to control the movement of the robot as well as its gripper. The

SVS classes have been used to interface the camera, and the BotSpeak library functions have been used to synthesize voice feedback from the robot. This chapter discusses the tools required for robot programming. This should facilitate better understanding of the codes that are presented in the subsequent chapters.

5.2 Robot Hardware and Software Resources

The Pioneer 2-DX ActivMedia mobile robot shown in Fig. 5.1, contains basic components for sensing and navigation in a real-world environment. It also includes battery power, drive motors and wheels, position–speed encoders, integrated sensors and accessories like a gripper and stereo camera. The robot is controlled by an onboard microcontroller and robot server software [Saphira, 1999]. Pioneer 2-DX also contains an addressable I/O bus for 16 devices, two RS-232 serial ports, eight digital I/O ports, and five A/D ports, which are accessible through a common application interface to the robot server software, Pioneer 2 Operating System (P2OS). The weight of Pioneer 2-DX is 9 kg and it can carry extra payload of up to 23 kg.

Fig. 5.1. Pioneer 2-DX from ActivMedia Robotic LLC, USA

5.2.1 Components

The main components of Pioneer 2-DX include deck(s) and console, body, nose, and accessory panels and an array of eight sonar sensors in the front and eight in the rear that provide object detection and navigation. The robot has reversible-DC motors equipped with a high-resolution optical quadrature shaft encoder for position and speed sensing and dead reckoning. The major component of the Pioneer 2-DX microcontroller is a 20 MHz Siemens 88C166 microprocessor with integrated 32K Flash-ROM and dynamic RAM of 32K. Information about the robot's state and connections appears on a 32-character (two lines) liquid-crystal display (LCD) on the console. The display shows the state of communication with the client computer along with the battery voltage. The RESET and MOTORS push-button switches on the console affect the microcontroller's logic and motor-driver systems. The serial communication between microcontroller and client can be established through the RS-232 port.

The Pioneer 2-DX is mounted with an onboard PC, hard-disk drive, onboard 10/100 Base-T Ethernet for network access along with 10-BaseT Ethernet cable to a 10/100 Mbps hub for network access. The operating system of Pioneer's onboard PC is LINUX 7.1, which includes networking features that support remote monitoring and control of the robot's onboard systems. Ethernet access is done by the Carrier Sense Multiple Access/ Collision Detection (CSMA/CD) technique and is established using BreezeNET PRO.11 Series.

The Small Vision System (SVS) is a built-in package for vision processing, which consists of several library functions for image capture and stereo correlation. A video-capture board or boards in the PC digitizes the video streams into main memory. The SVS library package computes the disparity image, which can be displayed or processed further. It also includes a program namely *smallv*, which is used for standalone vision processing applications.

5.3 ARIA

ARIA stands for ActivMedia Robotics Interface for Application, and was designed for use with ActivMedia Robotics mobile robots. The ARIA programming library is for C++ object-oriented programmers who want to have close control of the robot. ARIA is also useful for preparing robot-control software and deploying it on ActivMedia Robotics mobile robot platforms.

SRI International's Saphira [Konolige, 1995] has been built upon ARIA and is useful for creating applications with built-in advanced robotics capabilities, including gradient navigation and localization, as well as GUI controls with visual display of robot platform states and sensor readings. ARIA will be discussed in detail as it gives greater control for building programs to achieve desired results. The following sections cover the fundamental usage of ARIA for programming the mobile robot illustrated by a simple program whenever required. In detail the features of the package can be obtained from the ARIA product manual.

5.3.1 ARIA Client–Server

The mobile server is embodied in the Pioneer 2-DX operating system software and is found embedded on the robot's microcontroller, which manages the low-level tasks of robot control and operation, including motion, heading, as well as acquiring sensor information such as sonar, compass and driving accessories like the Pioneer gripper. The robot server does not perform robotic tasks, rather it is executed by the client. ARIA works mainly on the client side, which establishes and manages client–server communications between the ARIA-based software client and the robot's onboard servers. ArRobot is the heart of ARIA that acts as a client–server communication gateway and is responsible for the collection of state-reflection information such as position (x, y), velocity (translation and rotation) and heading/direction. It handles the client–server communication between applications software and the robot or the simulator as per the packet-based protocols. This functionality is provided by ArDevice-Connection, which is the base class to ArSerialConnection and ArTcpConnection, which are its built-in classes commonly used to manage communication between application software and the Pioneer 2-DX robot or SRIsim robot simulator respectively. ArSerialConnection opens a serial port and establishes a connection with the actual robot. ArTcpConnection connects to the robot or robot simulator through the network. In the case of the robot simulator SRIsim, the simulator listens on the 8101 port of the machine on which it is running. The ArTcpConnection class can be used to connect to the actual robot. To activate this before launching the client program, start ipthru on the robot (via telnet). This program acts as a bridge between the network and the serial port of the onboard computer of the robot.

After associating the device with the robot, it is required to establish the client–server connection between ARIA's ArRobot and the Pioneer 2-DX microcontroller or *SRIsim* simulator. The blockingConnect() method doesn't return from the call, until a connection succeeds or fails. ArRangeDevice are range device abstractions of the robot, for which there are relevant readings. ArSonarDevice has been used for handling sonar readings, which collect 2D data periodically at specific global coordinates. A RangeDevice is attached to the robot with

```
void ArRobot::addRangeDevice (ArRangeDevice
*device);
```

Sonar sensors are integrated with the robot controller and their readings are automatically included with the standard Server Information Packet (SIP) and are handled by the standard ArRobot packet handler. Nonetheless, it must be explicitly added to the sonar RangeDevice with the robot object to use the sonar readings for control tasks. Each RangeDevice has two sets of buffers (ArRangeBuffer): current and cumulative, and each supports two different reading formats: box and polar. The current buffer contains the most recent reading and the cumulative buffer contains several readings over time.

The ARIA client drives the robot and runs its various accessories through direct and motion commands, as well as through Actions. At the very lowest level, one may send commands directly to the robot server through ArRobot. Direct commands consist of a 1-byte command number followed by none or more arguments as defined by the robot's operating system. Motion commands are explicit movement commands and control the mobility of the robot, either to set an individual wheel or to coordinate translational and rotational velocities or to change the robot's absolute or relative heading or move a prescribed distance or just stop. The list of command functions of the robot used for motion commands is given in Table 5.1.

Table 5.1: List of command functions

Commands	Name of Function

mobility control	`void ArRobot::setVel2(double leftVel, double rightVel);`
	`void ArRobot::setVel(double velocity);`
	`//in mm/sec`
	`void ArRobot::setRotVel(double velocity); //in degrees/sec`

absolute and relative heading control	`Void ArRobot::setHeading(double heading) //in degrees` `ArRobot::setDeltaHeading(double delta)`
distance move	`Void ArRobot::move(double distance);`
Stop control	`void ArRobot::stop();`
check for completion of the motion command	`bool isMoveDone(double delta=0.0);` `bool isHeadingDone(double delta=0.0);`

Instead of using Direct or Motion commands, the ARIA client software uses `ArAction` to drive the robot. ARIA includes a number of built-in actions. Actions are added to robots with

```
ArRobot::addAction (ArAction *action, int prior-
ity),
```

including a priority which determines its position in the action list. Custom actions can be created by deriving from an abstract class namely,

```
ArAction. ArActionDesired * ArAc-
tion::fire(ArActionDesired currentDesired)
```

is the only function that needs to be overloaded for an action to work. Actions are evaluated by the resolving descending order of priority (lowest priority goes first) in each `ArRobot syncTask` cycle just prior to State Reflection. The resolver goes through the actions to find a single end `ActionDesired (ArActionDesired ())`, which defines the exact action to be performed. At the lowest level the robot's microcontroller manages the tasks of robot control and operation, which include movement of the mobile actuator and end-effectors such as motors and gripper, as well as acquiring sensor information such as sonar or IR beams. The program running on the microcontroller i.e. the operating system of the microcontroller, thus performs the task of mobile servers. The Pioneer 2DE operating system is one such OS that is embedded in the microcontroller of ActivMedia Robotics mobile robots. But the robot servers do not perform robotic tasks. They only perform elementary operations at the lowest level. To implement robotics control strategies and tasks, such as obstacle avoidance and detection, sensor fusion, feature recognition, map-building, intelligent navigation, etc., intelligent programs have to be run on a connected PC which interact with the servers. These client programs run the desired robotic tasks. An experiment has been conducted in the Laboratory with the following IP Address:

| Server | : 192. 168. 0. 1 |
| Robot | : 192. 168. 0. 9 |

The network setting for the client–server architecture of the robot is as follows, which will be used in the rest of the programs discussed in the subsequent chapters.

Hostname	: p2.local.net
IP	: 192.168.0.9
Netmask	: 255.255.255.0
Default Gateway	: 192.168.0.1
Primary DNS	: 192.168.0.1

The following commands are used to establish a connection between client and robot server.

```
$ xhost +192.168.0.9 ( 192.168.0.9 being added to
the control list)
$ telnet 192.168.0.9
Red Hat LINUX release 7.1 (Seawolf)
Kernel 2.4.2-2.VSBC6 on an i586
login: guest
Last login: Sun Apr 21 15:45:54 from 192.168.0.3
 guest@p2 guest]$ export DISPLAY=192.168.0.5:0.0
 guest@p2 guest] saphira
```

5.3.2 Robot Communication

The first task of any robotic application is to establish and manage client–server communication between the software client developed in ARIA and the onboard robot servers and devices. ARIA provides inbuilt support for this connectivity through its class `ArDeviceConnection`.

5.3.3 Opening the Connection

The code fragment given below shows how to establish connection to the simulator and the robot.

```
// serial connection (robot)
ArSerialConnection serConn;
// tcp connection (sim)
ArTcpConnection tcpConn;
// robot
ArRobot robot
Aria::init();
tcpConn.setPort("localhost",8101);
// tcpConn.setPort("192.168.0.9",8101);
// see if we can get to the simulator(true is suc-
cess)
if (tcpConn.openSimple())
{
printf("Connecting to simulator through tcp.\n");
robot.setDeviceConnection(&tcpConn);
}
else
{
serConn.setPort();//default "/dev/ttyS0" or "COM1"
serial port
printf("Could not connect to simulator, connecting
to robot through serial.\n");
robot.setDeviceConnection(&serConn);
}
```

The `Aria::init()` function initializes the ARIA system and is a must before any features of ARIA are used in the program. The program should call `Aria::shutdown()` to un-initialize or close the ARIA system. The instance 'robot' of `ArRobot` is the program's abstraction of the physical robot and is the most important instance in any program using which all robotic tasks shall be carried out. The instances of `ArSerial-Connection` and `ArTcpConnection` (in this case, `serConn` and `tcpConn`, respectively) are used to establish a connection. The above order ensures that the programs first try to connect to the simulator and if the simulator is unavailable it tries to connect to the robot server. The `set-Port()` function sets the `port_id` through which the communication is to take place. If such a communication port is successfully opened then a client–server connection is established between the program (i.e. the instance of `ArRobot`) and the onboard robot server, using the `setDeviceConnection()` function of the `ArRobot` class. In order to ensure that a non-blocking connection has been established the following lines of code are used.

```
if (!robot.blockingConnect())
{printf("Could not connect to robot... Exiting.");
Aria::shutdown();
return(1);
}
```

The `blockingConnect()` method doesn't return from the call until a connection succeeds or fails.

5.3.4 ArRobot

`ArRobot` as has been stated earlier, is an abstraction of the real robot, i.e. it is the program version of the physical robot. Any function called on the instance of the `ArRobot` class applies to the robot's servers. That means once the robot connection is established, instructions given to the ArRobot instance are reflected in the robot and the feedback or state of the robot is reflected as data in certain functions of the `ArRobot` class. It is for this reason that `ArRobot` is called the heart of ARIA because it acts as the client–server communications gateway and the central point for collection of state-reflection information such as position (x, y), velocity (translation and rotation), and heading/direction.

Furthermore, `ArRobot` also performs the important task of managing the program clockcycles and control of multithreading. `ArRobot` locks the clock cycle with the robot information packet cycle and performs synchronous tasks such as the Server Information Packet (SIP) handler, sensor interpreters, action handlers, state reflectors, user tasks and many more.

The state reflector functions are listed below. These enable the program to obtain the instantaneous state of the robot in terms of its position, velocity, heading, etc.

Std::string	**getRobotName** (void)	Returns the robot's name.
Std::string	**getRobotType** (void)	Returns the type of the robot connected to.
Std::string	**getRobotSubType** (void)	Returns the subtype of the robot connected to.
double	**getMaxTransVel** (void)	Gets the robot's maximum translational velocity.
double	**getMaxRotVel** (void)	Gets the robot's maximum rotational velocity.
bool	**setMaxRotVel** (double my-MaxVel)	Sets the robot's maximum rotational velocity.
ArPose	**getPose** (void)	Gets the global position of the robot.
double	**getX** (void)	Gets the global X location of the robot.
double	**getY** (void)	Gets the global Y location of the robot.
double	**getTh** (void)	Gets the global Th location of the robot.
double	**getVel** (void)	Gets the translational velocity of the robot.
double	**getRotVel** (void)	Gets the rotational velocity of the robot.
double	**getRobotRadius**(void)	Gets the robot radius (in mm).
double	**getRobotDiagonal** (void)	Gets the robot diagonal (half-height to diagonal of octagon) (in mm).
double	**getBatteryVoltage**(void)	Gets the battery voltage of the robot.
double	**getLeftVel** (void)	Gets the velocity of the left wheel.
double	**getRightVel** (void)	Gets the velocity of the right wheel.
Int	**getStallValue** (void)	Gets the 2 bytes of stall return from the robot.

5.3.5 Range Devices

Range devices (ArRangeDevice) are abstractions of the real robot sensors for which there are relevant readings. ArRangeDevice can represent any sensor that periodically collects 2D range data at specific global coordinates. Here we have used only ARIA RangeSensor: sonar. The sonar sensors are abstracted by class ArSonarDevice which is a child of the ArRangeDevice. Once a range device is instantiated using the desired parameters, it must be attached to the ArRobot instance of the program using the range device. This is done using the following function:

```
void ArRobot::addRangeDevice(ArRangeDevice
*device);
```

It is to be noted that sonar are integrated with the robot controller and that their readings automatically come included with the standard Server Information Packet (SIP) and so are handled by the standard ArRobot packet handler. Nonetheless, we must explicitly add the sonar RangeDevice with the robot object so that we can use the sonar readings for control tasks. Each RangeDevice has two sets of buffers (ArRangeBuffer): current and cumulative, and each support two different reading formats: box and polar

The function prototypes for both types of buffers for both the formats are as shown:

```
double ArRangeDevice::currentReadingPolar (double startAngle,
double endAngle, double *angle=NULL)

double ArRangeDevice::currentReadingBox (double x1, double
y1, double x2, double y2, ArPose *readingPos=NULL)

double ArRangeDevice::cumulativeReadingPolar (double startAngle, double endAngle, double *angle=NULL)

double ArRangeDevice::cumulativeReadingBox (double x1, double
y1, double x2, double y2, ArPose *readingPos=NULL)
```

In polar functions the return value is the nearest range within the sector formed between the startAngle and endAngle in the counterclockwise direction. In Box functions the return value is the nearest range in the rectangular box formed by the coordinates $x1, y1$ and $x2, y2$. The parameter angle is a pointer to a double value which gives the angle to the shortest range that the function returns. The current buffer contains the most recent reading; the cumulative buffer contains several readings over time. As stated earlier, the sonar readings are included in the SIP and can therefore be obtained using `ArRobot`. The function prototypes are as follows:

```
int ArRobot::getSonarRange(int num)
ArSensorReading * ArRobot::getSonarReading(int num)
```

The former returns the closest range returned by the sonar numbered by the parameter `num` while the latter returns a pointer to a `ArSensorReading` object for the sonar numbered `num`.

5.3.6 Commands and Actions

The ARIA client drives the robot and runs its various accessories through Direct and Motion commands, as well as through Actions.

Direct commands: At the very lowest level, one may send commands directly to the robot server through `ArRobot`. These commands are defined by the robot's operating system and consist of a 1-byte command number followed by none or more arguments. Direct commands to the robot come in five flavors, each defined by its command argument type and length:

```
ArRobot::com (unsigned char command)
Sends the command number without any arguments.

ArRobot::comInt (unsigned char command, short int arg)
Sends the command number with one byte argument

ArRobot::com2Bytes (unsigned char command, char high,
char low) Sends the command number with two bytes as
arguments.

ArRobot::comStr(unsigned char command, std::string arg)
Sends the command number with a string as argument.
```

The instruction

```
ArRobot::comStrN (unsigned char command, const char
*str, int size)
```

sends the command number with a character array and its size as arguments. For details refer to the ARIA User Manual.

Motion commands: These are explicit movement commands and act to immediately control the mobility of the robot. These are listed below, along with their purposes. To set an individual wheel, or coordinated translational and rotational velocities:

```
void ArRobot::setVel2(double leftVel, double rightVel);
void ArRobot::setVel(double velocity); //in mm/sec
void  ArRobot::setRotVel(double  velocity);  //in  de-
grees/sec
```

To change the robot's absolute or relative heading:

```
void  ArRobot::setHeading(double  heading)  //absolute
heading in degrees
ArRobot::setDeltaHeading(double delta)
```

To move a prescribed distance:

```
void ArRobot::move(double distance);
```

To stop the robot:

```
void ArRobot::stop();
```

The following functions check whether the previous motion command has been completed:

```
bool isMoveDone(double delta=0.0);
bool isHeadingDone(double delta=0.0);
```

A Direct or a Motion command is executed asynchronously, i.e. in the programs own thread and therefore has no coordination with the robot clock cycle. As a result, calls to Direct or Motion commands may conflict with controls from Actions or other upper-level processes and lead to unexpected consequences. For this purpose Direct motion commands are executed prior to Actions by giving them a precedence time using the function `void ArRobot::setDirectMotionPrecedenceTime` (unsigned int time)

If the time is set to 0 then a call to the function `void ArRobot::clearDirectMotion()` should be used to cancel the overriding effect of a Motion command so that Actions are able to regain control over the robot.

Actions: Actions are synchronously running threads which control the mobility of the robot and its accessories. As they run synchronously with the robots clock cycle, they can run in coordination with the instantaneous SIPs which contain essential parameters which can be used to modulate the movement of the robot. Actions are therefore useful to impart behavior to the robot which characterizes a number of robotic tasks

Hence, instesd of using Direct or Motion commands, it is preferable that the ARIA client software uses Actions to drive the robot. ARIA provides some predefined builtin Actions, all of which derive from the base class `ArAction`. These include `ArActionAvoidFront`, `ArActionAvoidSide`, `ArActionBumpers`, `ArActionConstantVelocity`, `ArActionStallRecover`, `ArActionGoto`, and many others. One may create an Action with desired properties by inheriting a class from the base class `ArAction`. `ArAction` defines a member function `fire()` which entirely specifies the nature of the Action.

```
ArActionDesired * ArAction::fire(ArActionDesired cur-
rentDesired)=0
```

This function needs to be overridden for an action. In specifying what the Action needs to do, the `fire()` function creates an instance of `ArActionDesired`. Actions are added to the robot's list of synchronous tasks by the following function:

```
ArRobot::addAction (ArAction *action, int prority)
```

The priority parameter enables `ArResolver` to resolve between two competing Actions and accordingly allot CPU time. The function `ArAction::setRobot (ArRobot *robot)` is called on an Action when it is added to a robot. Actions are evaluated on descending order of priority i.e. lowest priority goes first in each `ArRobot syncTask` cycle just prior to State Reflection. The resolver goes through the actions to find a single end `actionDesired` (`ArActionDesired` ( )). A number of competing actions determine the final motion commands that will be transmitted to the robot server. The ARIA library uses the fuzzy values of the competing actions to compute the final fuzzy value which is then defuzzified and sent to the robot's server where the final motion takes place.

The following program is a simple example of user-defined Action programming.

```
Source listing of sonar.cpp
// sonar.cpp
#include "Aria.h"
class ActionSonar : public ArAction
{
public:
// constructor, sets myMaxSpeed and myStopDistance
ActionSonar();
// destructor, which is just empty, nothing to be done
here
virtual ~ActionSonar(void) {};
// fire  is what the resolver calls to figure out what
this action wants
virtual ArActionDesired *fire(ArActionDesired current-
Desired);
// sets the robot pointer, also gets the sonar device
virtual void setRobot(ArRobot *robot);
protected:
// this is to hold the sonar device form the robot
ArRangeDevice *mySonar;
// what the action wants to do
ArActionDesired myDesired;
};
ActionSonar::ActionSonar() :
ArAction("Sonar")
{
mySonar = NULL;
}

/*
Sets the myRobot pointer (all setRobot overloaded func-
tions must do this), finds the sonar device from the
robot, and if the sonar isn't there, then it deacti-
vates itself.
*/
void ActionSonar::setRobot(ArRobot *robot)
```

```
{
myRobot = robot;
mySonar = myRobot->findRangeDevice("sonar");
if (mySonar == NULL)
deactivate();
}
//This fire is the whole point of the action.

ArActionDesired *ActionSonar::fire(ArActionDesired cur-
rentDesired)
{
double range;
double angle;
int num;
// reset the actionDesired (must be done)
myDesired.reset();
myDesired.setVel(0.0);
// if the sonar is null, nothing can be done and there-
fore deactivate
if (mySonar == NULL)
{
deactivate();
return NULL;
}
// get the range off the sonar
num=myRobot->getNumSonar();

range = mySonar->currentReadingPolar(-60, 60,&angle)
- myRobot->getRobotRadius();

// if the range is greater than the stop distance, find
some speed to go
// return a pointer to the actionDesired, so that re-
solver knows what to do

if (range>0)
{
printf("Range = %.2f and angle = %.2f ",range,angle);
}
if ((angle>=0.0)&&(angle<20.0)) printf("Left 1");
else if ((angle>=20.0)&&(angle<40.0)) printf("Left 2");
else if ((angle>=40.0)&&(angle<60.0)) printf("Left 3");
else   if   ((angle>=-20.0)&&(angle<0.0))   printf("Right
1");
else   if   ((angle>=-40.0)&&(angle<-20.0))   printf("Right
2");
else   if   ((angle>=-60.0)&&(angle<-40.0))   printf("Right
3");
```

```
ArSensorReading* r=myRobot->getSonarReading(0);
printf("(%.2f,%.2f)%.2f.\n",r->getX(),r->getY(),r->getSensorTh());

return &myDesired;
}

int main(void)
{
// The connection is used to talk to the robot
ArSerialConnection con;
// the robot is defined
ArRobot robot;
// the sonar device is defined
ArSonarDevice sonar;

// some stuff for return values
int ret;
std::string str;

// the behaviors from above, and a stallRecover behav-
ior that uses defaults
ActionSonar asonar;

// this needs to be done
Aria::init();

 // open the connection using the defaults, if it
fails, exit
if ((ret = con.open()) != 0)
{
str = con.getOpenMessage(ret);
printf("Open failed: %s\n", str.c_str());
Aria::shutdown();
return 1;
}

// add the range device to the robot,
//you should add all the range devices before you add
actions
robot.addRangeDevice(&sonar);
// set the robot to use the given connection
robot.setDeviceConnection(&con);

// do a blocking connect, if it fails exit
if (!robot.blockingConnect())
```

```
{
printf("Could not connect to robot... exiting\n");
Aria::shutdown();
return 1;
   }
// enable the motors, disable amigobot sounds
robot.comInt(ArCommands::ENABLE, 0);
robot.comInt(ArCommands::SOUNDTOG, 0);

// add actions in order, the integer is used here for
the priority, with higher priority actions going first

robot.addAction(&asonar, 50);

// runs the robot, the true is used to exit, if it
loses connection
robot.run(true);

// used to shutdown and go away
Aria::shutdown();
return 0;
}
```

Compile and link
gcc –c –I$ARIA/include sonar.cpp
gcc –o sonar –L$ARIA/lib –lAria –ldl –pthread sonar.o

Saphira: Saphira is a mobile robotics-client development environment, which is a product of SRI International's Artificial Intelligence Center [Konolige, 1995] and operates in a multitiered client–server environment. Saphira carries the basic components of real-world sensing and navigation for intelligent mobile robot activities, including drive motors and range-finding sensors as well as embedded controllers to manage various resources. It handles the low-level details of sensor and drive management such as collection of range-finding information from onboard sonars, positioning, heading, and so on. Saphira provides the intelligence for various operations of the robot server. Saphira's lowest level is interfaced with the robot that provides a coherent method and protocols for communication and control of the robot server. Saphira's intermediate layers support higher-level functions for navigation control and sensor interpretation. Saphira also provides a Graphical User Interface (GUI) and command-level interface through Colbert Executive for interactive monitoring and manual control of both Saphira client and robot server with accessories.

Colbert has two processes, namely finite state automata (FSA) and concurrent processes. A program in Colbert is an *activity* whose semantics are based on FSA. *Activity* controls the overall behavior of the robot in several ways, such as

- sequencing the basic actions that the robot performs;
- monitoring the execution of basic actions and other activities;
- executing activity subroutines;
- checking and setting the values of internal variables.

FSA are used to reason about computational complexity and decidability. The advantage of Colbert lies in its ability to make an intuitive translation from operator constructs in ANSI C to FSA capable of controlling the robot.

5.4 Socket Programming

A bi-directional communication device is used to communicate with another process on the same machine or with a process running on other machines. Sockets are the only interprocess communications that are used to permit communication among the different computers and the robot's server [Rubini, 1998]. A socket may be viewed as a software association with a hardware port from which the processes read from or write to. In LINUX, a socket is created by associating a reference to a `port id` in each of the communicating programs with the following parameters: communication style, namespace and protocol. Data are sent and received in packed chunks called packets. The communication style determines how these packets are handled and how they are addressed. Namespace specifies how the socket addresses will be written and the protocol determines how data are transmitted. The protocols used in socket programming are TCP/IP, the APPLETALK network protocol, or UNIX local communication protocol. The above functionality is usually wrapped into methods of a class Socket to provide a ready-to-use simple communication format for other programs. JAVA makes things easier by providing a standard API library and has built-in support for socket programming.

In a client–server architecture, a socket may be configured to be a server socket or a client socket, the difference being that a server socket continuously checks (listens) for a connect request from a client, while the client socket actually initiates the connection request. The server socket then validates and accepts the client connection following which interprocess communication begins.

Here we are concerned with implementing a network interface in our custom programs. This is achieved by creating programs with client–server interfacing, which means each program consists of two different parts, i.e. the server side and the client side. Here one must not be confused with the ARIA client–server concept discussed in the preceding sections. While the robot servers 'serve' the ARIA client program which executes robotic tasks, at the same time the ARIA client itself acts as a server for a remote machine connected through a radio network, where client's programs are used as remote control front ends. Front ends are discussed in a subsequent chapter. Here we will discuss the concepts of developing the network server programs using socket programming.

The server: The server is the program running on the robot and it offers all services to a client. Usually these may be instantaneous robot parameters, environment characteristics such as sonar readings, images from the camera, etc., depending on the application. Server programs have been developed in C++, with the strong support of available libraries for convenience of coding.

The client: The client program, written in JAVA, runs on a local machine and it sends requests or commands to the server either to execute a routine or to procure data as mentioned earlier.

The socket-programming concept has been used in the design of the client–server architecture. A socket is a logical reference to hardware or a software port on to which programs can write or read data. Such a socket is wrapped with formatting functions for the purpose of communicating formatted data instead of binary data. A server creates a socket and listens, i.e. it waits till a client requests connections. On reception of a request, it accepts the client by acknowledging the client connection.

5.4.1 Socket Programming in ARIA

ArSocket is a wrapper, around the socket network layer, which allows clients to use the socket's networking interface in an operating system independently using standard socket functions. This class also contains the file descriptor which identifies the socket to the operating system. In the Windows operating system the networking subsystem needs to be initialized and shut down individually by each program. Therefore, a program starts by calling the static function `ArSocket::init()` and exit by calling `ArSocket::shutdown()`. For programs that use `Aria::init()` and `Aria::uninit()`, calling `ArSocket::init()` and `ArSocket::shutdown()` is unnecessary. The ARIA initialization functions take care of this and these functions do nothing in LINUX.

A server socket may be created by using the following constructor:

ArSocket::ArSocket (int *port*, Type *type*),
where *port* is the port id and *type* is the type of protocol to be used and is usually `ArSocket::TCP`

Alternatively one may call the following function:

`ArSocket::open(int port,Type type);`

This server socket can be used to accept a client by using the following function:

`ArSocket::accept(ArSocket *clientSocket);`

The `accept` function listens at the port id for a client to request for connection and gets a pointer (`ArSocekt*`) to the remote client for use in the program. All inputs and outputs are now directed from and to `clientSocket`. To connect to a server socket one has to use the constructor:

`ArSocket::ArSocket ( const char * host, int port, Type type)`

It connects to the socket at a given port id on the machine specified by the host id. Alternatively one may use

`ArSocket::connect(const char * host, int port,Type type);`

One may read and write data through the socket using the following functions :

`size_t ArSocket::read ( void * buff, size_t len, unsigned int msWait = 0) [inline]`

Read from the socket

Parameters:

```
buff          buffer to read into
len           how many bytes to read
msWait        if  0,  don't  block,  if  >  0  wait  this
              long for data
Returns:      number of bytes read
```

```
size_t  ArSocket::write  ( const  void  *      buff,
size_t    len) [inline]
```

Write to the socket.

Parameters:

```
buff          buffer to write from
len           how many bytes to write
```

```
Returns:      number of bytes written
```

5.5 BotSpeak Speech System

BotSpeak is speech-synthesis, speech-recognition software integrated for use with Pioneer Intelligent Mobile Robots. It is basically a C library and provides a set of predefined functions which can be used for speech-synthesis or recognition. Here we will briefly cover a few functions used for speech synthesis. For speech recognition the BotSpeak server defines contexts for each program and words present in the contexts are recognized. Speech-recognition functions are not covered here as they have not been used in the programs.

5.5.1 Functions

```
void bsInit(void)
```

This function starts the ViaVoice/BotSpeak server if it is not already running and initializes BotSpeak. If the server running from a previous call to bsInit(), it connects BotSpeak automatically. This function also clears out any words or contexts that the server already had, which means only one program can run at a time when connected to the server. This is designed in such a way that when one restarts a client program, the program will start in a known state without any ambiguities created due to contexts defined by other programs.

```
void bsSpeak(char *string)
```

This function passes the string argument to the ViaVoice/BotSpeak server to synthesize voice and pronounce or utter the string of characters. The microphone will be turned off while synthesis takes place. This function can also be used to change the voice qualities by sending a certain predefined sequence of characters. Some of these are listed below.

`vbN   Pitch, N in range 0 to 100
`vhN   Head size, N in range 0 (tiny head) to 100 (huge head)
`vrN   Roughness, N in range 0 (smooth) to 100 (rough)
`vyN   Breathiness, N in range 0 (not breathy) to 100 (breathy whisper)
`vfN   Pitch fluctuation, N in range 0 (monotone) to 100 (wide fluctuation)
`vsN   Speed, N in range 0 (slow) to 250 (Fast)
`vvN   Volume, N in range 0 (soft) to 100 (loud)
`vg0   Set voice to male
`vg1   Set voice to female
`00   Reduced emphasis
`0   No emphasis
`1   Normal emphasis
`2   Added emphasis
`3   Heavy emphasis
`4   Very heavy emphasis

Note that if the function is called while there is no playing or synthesis going on the function will return immediately, but if synthesis or playing is occurring it will pause before returning (until at least some of the previous synthesis is done).

```
void bsFinishSpeaking(void)
```

This function waits until the previous speech synthesis is completed i.e. this function will return when BotSpeak has finished with the previous `bsSpeak()` or `bsPlay()` (`bsPlay()` is called for playing an audio file). The following program illustrates how to use BotSpeak in a C++ program for speech synthesis.

```cpp
//BotSpeak.cpp
//BotSpeak sample program
#include <stdio.h>
#include <stdlib.h>
#include <string.h>
extern "C"{
#include "BotSpeak.h"
}
int main(int argc, char **argv)
{
char text[121];
// initialize botspeak, if the server isn't run-
ning this starts it
bsInit();
// say something so people know we're alive
bsSpeak("Type any sequence of words and I shall
say it.");
bsFinishSpeaking();

while(1)
{
printf("Enter text:-");
fgets(text,120,stdin);
printf("saying: %s",text);
bsSpeak(text);
bsFinishSpeaking();
if (strcmp(text,"exit\n")==0) break;
}
return(0);
}
```

Compilation and linking in LINUX
gcc -c -I$ BotSpeak /include botspeak.cpp
gcc -o botspeak -L$BOTSPEAK/lib -lbotspeak botpeak.o

5.6 Small Vision System (SVS)

The Small Vision System (SVS) is meant to be a development environment for applications requiring stereo processing. It consists of a library of functions for performing image capture and stereo correlation. Images come in via a pair of aligned video cameras, called a stereo head. A video-capture board or boards in the PC digitizes the video streams into main memory. The SVS functions are then invoked, given a stereo pair as an argument. These functions compute a disparity image, which the user can display or process further. The SVS provides library functions for C that are defined under svs.h. However for C++ programmers built-in classes are available for ease of use under svsclasses.h. We will briefly discuss the classes that are available for image capture and processing in C++.

5.6.1 SVS C++ Classes

There are three main classes for SVS: classes that encapsulate stereo images, classes that produce the images from video or file sources, and classes that operate on stereo images to create disparity and 3D images. These classes are declared in the header file src/svsclass.h. The basic idea is to have one class (svsStereoImage) for stereo images and the resultant disparity images, which performs all necessary storage allocation and insulates the user from having to worry about these issues. Stereo image objects are produced from video sources, stored image files, or memory buffers by the svsAcquireImages classes, which are also responsible for rectifying the images according to parameters produced by the calibration routines. Disparity images and 3D point clouds are produced by the stereo processing class svsStereoProcess acting on a stereo image object, with the results stored back in the stereo image object. The basic operations are:

1. Make a video source object and open it. Which video source is used depends on which framegrabber interface file has been loaded.
2. Open the video source.
3. Set the frame size and any other video parameters you wish, and read in rectification parameters from a file.
4. Start the video acquisition.
5. Loop:

a. Get the next stereo image.
b. Do various processing required on the images.
c. Display the results.

The different classes that are defined under svsclasses.h are elaborated as follow.

5.6.2 Parameter Classes

svsImageParams: Image frame size and subwindow parameters
svsRectParams: Image rectification parameters
svsDistParams: Image stereo processing (disparity) parameters

Parameter classes contain information about the format or processing characteristics of stereo image objects. Each stereo image object contains an instance of each of the above classes. Application programs can read these parameters to check on the state of processing or the size of images, and can set some of the parameters, either directly or through class member functions.

Class svsImageParams: Frame size and subwindow parameters for stereo images. In general, the only way these parameters should be changed is through member functions of the appropriate objects, e.g. using SetSize in the svsVideoImages class.

Class svsRectParams: Rectification parameters for stereo images. They are used internally by the rectification functions. Application programs should not change these parameters, and will have few reasons to look at the parameter values. Rectification parameters are generated initially by the calibration procedure, then written to and read from parameter files.

Class svsDistParams: Disparity parameters control the operation of stereo processing, by specifying the number of disparities, whether left/right filtering is on, and so on. Most of these parameters can be modified by application programs.

5.6.3 Stereo Image Class

svsStereoImage: Stereo image class

The stereo image class encapsulates information and data for a single stereo image pair, along with any of its processed results, e.g., disparity image or 3D point cloud. Stereo image objects are usually produced by one of the image acquisition classes (svsVideoImages or svsFileImages), then processed further by an svsStereoProcess object. An svsStereoImage object holds information about its own state. For instance, there are Boolean flags to tell if there is a valid set of stereo images, whether they are rectified or not, if a valid disparity image has been computed, and so on. The svsStereoImage class handles all necessary allocation of buffer space for images. User programs can access the image buffers, but should be careful not to de-allocate them or destroy them

Constructor and Destructor

```
svsStereoImage();
~svsStereoImage();
```

Constructor and destructor for the class. The constructor initializes most image parameters to default values, and sets all image data to NULL.

```
char error[256];
```

If a member function fails (e.g. if ReadFromFile returns false), then the error will usually contain an error message that can be printed or displayed.

Stereo Images and Parameters

```
bool haveImages;// true if good stereo images have
been captured
bool haveColor;// true if left image color array
present
bool haveColorRight;// true if right image color
array present
svsImageParams ip;// image format, particular to
each object
unsigned char *Left();// left image array
unsigned char *Right();// right image array
unsigned long *Color();// left-color image array
unsigned long *ColorRight();// right-color image
array
```

These members describe the stereo images present in the object. If stereo images are present, haveImages is true. The stereo images are always monochrome images, 8 bits per pixel. Additionally, there may be a color image, corresponding to the left image, if requested. Color images are in RGBX format (32 bits per pixel, first byte red, second green, third blue, and fourth undefined). If the left color image is present, haveColor is true. The color image isn't used by the stereo algorithms, but can be used in post-processing, for example, in assigning color values to 3D points for display in an OpenGL window. Similarly, if the right color image is present, `haveColorRight` is true. The color images may be input independently of each other. Frame size parameters for the images are stored in the variable `ip`. The parameters should be considered read-only, with one exception: just before calling the `SetImage` function. The `Left`, `Right`, and `Color` functions return pointers to the image arrays. User programs should not delete this array, since the stereo object manages it.

Rectification Information

```
bool  isRectified;//whether  the  rectification  has
been done already
bool haveRect;//true  if  the  rectification  params
exist
svsRectParams  rp;//rectification  params,  if  they
exist
```

The images contained in a stereo image object (left, right and left-color) can be *rectified*, that is, corrected for intra-image (lens distortions) and inter-image (spatial offset) imperfections. If the images are rectified, then the variable `isRectified` will be true. Rectification takes place in the `svsAcquireImage` classes, which can produce rectified images using the rectification parameters. The rectification parameters can be carried along with the stereo image object, where they are useful in further processing, for example, in converting disparity images into a 3D point cloud. If rectification parameters are present, the `haveRect` variable is true. The rectification parameters themselves are in the `rp` variable.

Disparity Image

```
bool haveDisparity; // whether the disparity image
has been calculated yet
svsDisparityParams dp; // disparity image parame-
ters
short *Disparity(); // returns the disparity image
```

The disparity image is computed from the stereo image pair by an `svsStereoProcess` object. It is an array of short integers (signed, 16 bits) in the same frame size as the input stereo images. The image size can be found in the `ip` variable. It is registered with the left stereo image, so that a disparity pixel at X, Y of the disparity image corresponds to the X, Y pixel of the left input image. Values −1 and −2 indicate the absence of any disparity information: −1 for low-texture areas, and −2 for disparities that fail the left/right check. If the disparity image has been calculated and is present, then haveDisparity is true. The parameters used to compute the disparity image (number of disparities, horopter offset, and so on) are in the parameter variable dp. The disparity image can be retrieved using the Disparity function. This function returns a pointer to the disparity array, so it is very efficient. User programs should not delete this array, since the stereo object manages it.

3D Point Array

```
bool have3D;// whether 3D information is available
int numPoints;// number of points actually found
float *X(), *Y(), *Z();// 3D point arrays
```

The 3D point arrays are the 3D points that correspond to each pixel in the left input image. It has the same size (width and height) as the input stereo images. The 3D point array is computed from the disparity image using the external camera calibration parameters stored in rp. An svsStereo-Process object must be used to compute it. Each point is represented by a coordinate (X, Y, Z) in a frame centered on the left camera focal point. The Z-dimension is the distance from the point perpendicular to the camera plane, and is always positive for valid disparity values. The X-axis is horizontal and the positive direction is to the right of the center of the image; the Y-axis is vertical and the positive direction is down relative to the center of the image (a right-handed coordinate system). Negative values of Z are used to indicate there was no valid disparity reading at a pixel. If the 3D array is present, then have3D is true. The actual number of 3D points present in the arrays is given by numPoints.

File I/O

```
bool SaveToFile(char *basename);  // saves images
and params to files
bool ReadFromFile(char *basename);  // gets images
and params from files
bool ReadParams(char *name);  // reads just params
from file
bool SaveParams(char *name);  // save just params
to file
```

Images and parameters in a stereo object can be saved to a set of files
(`SaveToFile`), and read back in from these files (`ReadFromFile`).
The basename is used to create a file set. For instance, if the basename is:

TESTIMAGE, then the files set is:
TESTIMAGE-L.bmp // left image, if present
TESTIMAGE-R.bmp // right image, if present
TESTIMAGE-C.bmp // left color image, if present
TESTIMAGE.ini // parameter file

Just the parameters can be read from and written to a parameter file, using
`ReadParams` and `SaveParams`.

5.6.4 Acquisition Classes

The list of acquisition classes that are used to get stereo image data are as
follows:

```
svsAcquireImages   Base class for all acquisition
svsVideoImages     Acquire from a video source
svsFileImages      Acquire from a file or memory
                   source
```

Acquisition classes are used to get stereo image data from video or file
sources, and put into `svsStereoImage` structures for further processing. During acquisition, images can be *rectified*, that is, put into a standard
form with distortions removed. Rectification takes place automatically if
the calibration parameters have been loaded into the acquisition class. The
two subclasses acquire images from different sources. `svsVideoI-
mages` uses the capture functions to acquire images from a video device
such as the MEGA-D stereo head. `svsFileImages` acquires images
from BMP files stored on disk.

Constructor and Destructor

```
svsAcquireImages();
virtual ~svsAcquireImages();
```

These functions are usually not called by themselves, but are implicitly called by the constructors for the subclasses `svsVideoImages` and `svsFileImages`.

Rectification

```
bool HaveRect();
bool SetRect(bool on);
bool GetRect();
bool IsRect();
bool ReadParams(char *name);
bool SaveParams(char *name);
```

These functions control the rectification of acquired images. `HaveRect()` is true when rectification parameters are present; the normal way to input them is to read them from a file, with `ReadParams()`. The argument is a file name, usually with the extension `.ini`. If the acquisition object has rectification parameters, they can be saved to a file using `SaveParams()`. Rectification of acquired images is performed automatically if `HaveRect()` is true, and rectification processing has been turned on with `SetRect()`. Calling `ReadParams()` will also turn on `SetRect()`. The state of rectification processing can be queried with `GetRect()`. If the current image held by the acquisition object is rectified, the `IsRect()` function will return true.

Controlling the Image Stream

```
bool CheckParams()
bool Start()
bool Stop()
svsStereoImage *GetImage(int ms)
```

An acquisition object acquires stereo images and returns them when requested. These functions control the image streaming process. `CheckParams()` determines if the current acquisition parameters are consistent, and returns true if so. This function is used in video acquisition, to determine if the video device supports the modes that have been set. `Start()` starts the acquisition streaming process. At this point, images are streamed into the object, and can be retrieved by calling `GetImage()`. `GetImage()` wait upto 30 milliseconds for a new image before it returns; if no image is available within this time, it returns NULL. If an image is available, it returns an svsStereoImage object containing the image, rectified if rectification is turned on. The svsStereoImage object is controlled by the acquisition object, and the user program should not delete it. The contents of the svsStereoImage object are valid until the next call to `GetImage()`. `Start()` returns false if the acquisition process cannot be started. `Stop()` will stop acquisition.

Error String

```
char *Error()
```

Call this function to get a string describing the latest error on the acquisition object. For instance, if video streaming could not be started, Error() will contain a description of the problem

Video Acquisition

The video acquisition classes are subclasses of `svsAcquireImages`. The general class is `svsVideoImages,` which is referenced by user programs. This class adds parameters and functions that are particular to controlling a video device, e.g., frame size, color mode, exposure, and so on. Particular types of `framegrabbers` and stereo heads have their own subclasses of `svsVideoImages`. In general, the user programs won't be aware of these subclasses, instead treating them as a general `svsVideoImage` object. To access the `svsVideoImages` object, the special function `svsGetVideoObject()` will return an appropriate object.

Video Object

```
svsVideoImages *svsGetVideoObject()
```

Returns a video acquisition object suitable for streaming video from a stereo device. The particular video object that is accessed depends on the video interface library that has been loaded. This function creates a new video object on each call; so several devices can be accessed simultaneously, if the hardware supports it.

Opening and Closing

```
bool Open(char *name = NULL)
bool Open(int devnum)
bool Close()
```

The device must be opened before capturing the images by the stereo device. The `Open()` call opens the device, returning true if the device is available. An optional name can be given to distinguish among several existing devices. The naming conventions for devices depend on the type of device; typically it is a serial number or other identifier. These identifiers are returned by the `Enumerate()` call.

Alternatively, a number can be used, giving the device in the order returned by the `Enumerate()` function, i.e. 1 is the first device, 2 is the second, and so on. A value of 0 indicates any available device. Upon opening, the device characteristics are set to default values. To set values from a parameter file, use the `ReadParams()` function. A stereo device is closed and released by the `Close()` call.

Image Framing Parameters

```
bool SetSize(int w, int h)
bool SetSample(int decimation, int binning)
bool SetOffset(int ix, int iy, int verge)
bool SetColor(bool on, bool onr = false)
bool CheckParams()
```

These functions control the frame size and sampling mode of the acquired image. SetSize(w,h) sets the width and height of the image returned by the stereo device. In most cases, this is the full frame of the image. For instance, most analog frame grabbers perform hardware scaling, so that almost any size image can be requested, and the hardware scales the video information from the imager to fit that size. In most analog frame grabbers, the sampling parameters (decimation and binning) are not used, and a full-frame image is always returned, at a size given by the `SetSize()` function. Some stereo devices, such as the MEGA-D, allow the user to specify a subwindow within the image frame. The subwindow is given by a combination of sampling mode and window size. The sampling mode can be specified by `SetSample()`, which sets binning and decimation for the imager. The MEGA-D supports sub-sampling the image at every 1, 2 or 4 pixels; it also supports binning (averaging) of 1 or 2 (a 2 × 2 square of pixels is averaged). For example, with binning = 2 and decimation = 2, the full frame size is 320 × 240 pixels. Using `SetSize()`, a smaller sub-window can be returned. The offset of the subwindow within the full frame comes from the `SetOffset()` function, which specifies the upper left corner of the subwindow, as well as a *vergence* between the left and right images. `SetColor()` turns color on the left image on or off. Additionally, some applications require color from the right imager also, and setting the second argument to true will return a color image for the right imager. Generally, returning color requires more bus bandwidth and processing, so use color only if necessary.

The video frame parameters can be set independently, and not all combinations of values are legal. The `CheckParams()` function returns true if the current parameters are consistent. None of the frame or sampling mode parameters can be changed while images are being acquired, except for the offset parameters. These can be changed at any time, to pan and tilt the subwindow during acquisition.

Image Quality Parameters

```
bool  SetExposure(bool  auto,  int  exposure,  int
gain)
bool SetBalance(bool auto, int red, int blue)
bool SetLevel(bool auto, int brightness, int con-
trast)
```

These functions set various video controls for the quality of the image, including color information, exposure and gain, brightness and contrast. Not all stereo devices support all of the various video modes described by these parameters. In general, parameters are normalized to be integers in the range [0,100]. `SetExposure()` sets the exposure and gain levels for the device. If auto is chosen, the manual parameters are ignored. `SetBalance()` sets the color balance for the device. Manual parameters for red and blue differential gains are between −40 and 40. If auto is chosen, the manual parameters are ignored. `SetLevel()` sets the brightness and contrast for the device. In auto mode, the brightness value is ignored. Contrast is always set manually. These functions can be called during video streaming, and their effect is immediate. The following program illustrates how to obtain an image and store it in a file.

```
// imagesave.cpp
#include<stdio.h>
#include<stdlib.h>
```

```
#include<string.h>
#include<math.h>
#include<string>
#include"svsclass.h"
#define H 240
#define W 320

svsVideoImages *videoObject; // source of images
svsStereoImage *imageObject;
char filename[50],name[50];
char color_image[4*W*H+104];
int main(int argc, char **argv)
{
printf("\nEnter file name :");
scanf("%s",name);
videoObject = getVideoObject();
//size
svsImageParams *ip=videoObject->GetIP();
ip->width=W;
ip->height=H;
//sample
videoObject->decimation=2;
videoObject->binning=2;
svsHasMMX=true;
ip->linelen=ip->width;
ip->lines=ip->height;
ip->vergence=0;
bool ret;
videoObject->ReadParams("/root/megad-75.ini");
ret = videoObject->Open(0);
if (!ret) {printf("Can't open frame grabber.\n"); re-
turn 0; }
else
printf("Opened frame grabber.\n");
if (!videoObject->CheckParams())
{
printf("Incorrect  Params\n");
videoObject->Close();
exit(1);
}
videoObject->Start();
videoObject->SetColor(true);
videoObject->exposure=100;
videoObject->gain=95;   //30;   //Color 30
videoObject->blue=20;
videoObject->red=20;
videoObject->SetDigitization();
//Image grabber up and running
```

```
videoObject->Start();
for(int i=0;i<5;i++) imageObject = videoObject-
>GetImage(400);
if (imageObject != NULL)
{      sprintf(filename,"/home/images/%s",name);
if(imageObject->haveImages)
{
 memcpy(color_image,imageObject->color,4*W*H);
 imageObject->SaveToFile(filename);sendImage();
}
videoObject->Stop();
 }
printf("Done , exiting\n");
videoObject->Close();
imageObject->Close();
// shutdown
Aria::shutdown();
return(0);
}


/*
For Compilation  and Linking
g++ -c -I$SVS/src imagesave.cpp
g++ -o imagesave -L$SVS/bin -ldl -lcap -lsvs -pthread
sampler.o
*/
```

5.7 Multithreading

Multithreading is a conceptual programming technique. The program or the process is subdivided into two or more subprograms, which are implemented at the same time in parallel. Each flow of control is a separate tiny program known as a thread, which runs independently except when two or more threads compete for the same resources and share a part of the CPU time with other threads. The term parallel could be misleading as there is no true parallel execution, but only a time-shared operation where the CPU switches at random from one thread to another after a certain time. Thus threads can be either waiting, active or dead/suspended. The CPU time conflict is usually resolved by setting priorities and the thread with the highest priority gets the largest chunk of the CPU time.

Multithreading is usually implemented by defining a function `run()` in each of the routines to be run in parallel and then defining the thread body within it. The CPU uses these run routines to implement multithreading. The same principle is also made use of in achieving synchronized motion of the robots by means of Actions. Actions are nothing but synchronously running motion command threads which can allow the robot to synchronize instantaneous state variables with motion. The `run()` method is the only method in which the thread's behavior can be implemented.

5.8 Client Front-End Design Using JAVA

The client front-end design is very useful in a multiplatform network. The design can help the client to add various modules in the same program and test their different algorithms with minimum modifications, which releases the burden to design the front-end. Here, the program execution may be controlled and monitored by text input, but it is always advantageous to have a display which emulates a real-life control panel pertaining to the specific needs of the task. Therefore the Graphical User Interface (GUI) front-end gains importance for better representation and ease of control. Adjustable control components can be added to provide means for interactive control of run-time parameters. Examples of such components are choice boxes, text fields, buttons, scroll bars, etc.

GUI design may be accomplished in different ways using various programming languages such as Visual Basic, C++, JAVA, Delphi, etc. Among these front-end design tools, JAVA is chosen to be the best because of its platform independence and it has built-in support for most commonly used GUI components. However, it gives the programmer the power to build custom components by inheriting and adding properties to the existent components. In subsequent chapters we will discuss different client–server programs using JAVA.

5.9 Summary

This chapter gives essential ingredients for readers to develop the various application programs on robots using ARIA libraries.

6 Robot Parameter Display

6.1 Introduction

It is always essential to measure the robot state parameters such as its current position and heading, in which direction it is moving and others such as battery voltage. The robot parameters are used almost in every application of the robot, which helps during troubleshooting. Here a sample client –server program is given, whose objective is to collect all the robot parameters and display them within a frame on the client's screen. The client gets the information about the robot parameters from the server program running on the robot continuously and displays a frame showing all the parameters, such as the robot's name, its type, maximum translational and rotational velocity, current translational and rotational velocity, its battery voltage, its current position in rectangular coordinates, its heading, right and left wheel velocities etc. These parameters are generally monitored at the client, while the robot does some task in the environment. This helps the programmer to monitor and control the robot from a distance. The subsequent section elaborates the algorithm and client–server program to get the robot parameters and display them.

6.2 Flow Chart and Source Code for Robot Parameter Display

The client and server flow charts are given in Figs. 6.1 and 6.2. The server runs on the onboard computer where robot parameters are available and the client displays these parameters in a program window. The sample program given in Listing 6.1 illustrates the server program written in C++ and Listing 6.2 illustrates the client program written in JAVA. The program log session and output are shown in Fig. 6.3.

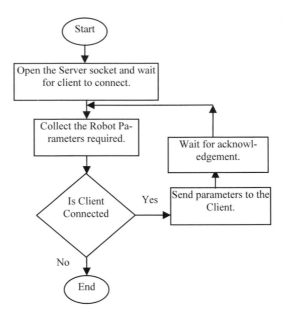

Fig. 6.1. Server program's flow chart

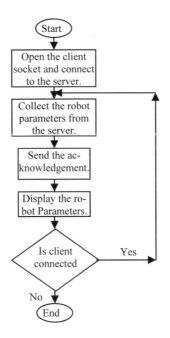

Fig. 6.2. Client program's flow chart

Listing 6.1. Server program

```cpp
//filename :  param.cpp
#include "Aria.h"
#include <string>

int main()
{
ArSerialConnection con;
ArRobot robot;

Aria::init(Aria::SIGHANDLE_THREAD);

if (con.open() != 0)
{
printf("Could not open the connection");
exit(1);
}

robot.setDeviceConnection(&con);

if (!robot.blockingConnect())
{
printf("\n Could not connect to robot");
Aria::shutdown();
exit(1);
}

// The string to send to the client. Done here as a
char array so that its
// easier to figure out its length.
// The buffer in which to recieve the hello from the
client
        char buff[100];

// The size of the string the client sent
size_t strSize;

// The socket object
ArSocket serverSock, clientSock;

// Initialize Aria. For Windows, this absolutely must
be done. Because
// Windows does not initialize the socket layer for
each program. Each
// program must initialize the sockets itself.
```

```
// Lets open the server socket
if (serverSock.open(7775, ArSocket::TCP))
printf("Opened the server port\n");
else
{
printf("Failed to open the server port: %s\n",
               serverSock.getErrorStr().c_str());
return(-1);
}

// Lets wait for the client to connect to us.
if (serverSock.accept(&clientSock))
   printf("Client has connected\n");
 else
printf("Error in accepting a connection from the cli-
ent: %s\n",
serverSock.getErrorStr().c_str());

// Lets send the string 'Hello Client' to the client.
The write should
// return the same number of bytes that we told it to
write. Otherwise,
// its an error condition.

std::string robtype = robot.getRobotType();
std::string robname = robot.getRobotName();

robot.runAsync(true);

char fixeddata[100];

sprintf(fixeddata,"%s|%s|%.01f|%.01f|%.01f|%.01f",
robname.c_str(),robtype.c_str(),robot.getRobotRadius(),
robot.getRobotDiagonal(),robot.getMaxTransVel(),
robot.getMaxRotVel());

clientSock.write(fixeddata,strlen(fixeddata));
clientSock.read(buff,sizeof(buff));

do{
char datastring[100];
robot.lock();
int a;
sprintf(datastring,"%.01f|%.01f|%.01f|%.01f|%.01f|%.01f
|%.1lf|",
robot.getX(),robot.getY(),robot.getTh(),robot.getVel(),
robot.getLeftVel(),robot.getRightVel(),
robot.getBatteryVoltage());
```

```
robot.unlock();

if((a=clientSock.write(datastring, strlen(datastring)))
                == strlen(datastring))
{}
else
{
printf("Error in sending hello string to the client
%d\n",a);
return(-1);
}

// Lets wait for the client to say hello to us.
strSize=clientSock.read(buff, sizeof(buff));
// If the amount read is 0 or less, its an error condi-
tion.
if (strSize > 0)
{
// Terminate the string with a NULL character.
     buff[strSize]='\0';
}
else
{
printf("Error in waiting/reading the hello from the
client\n");
return(-1);
}
}while(true);

// Now lets close the connection to the client
clientSock.close();
printf("Socket to client closed\n");
// And lets close the server port
serverSock.close();
printf("Server socket closed\n");

// Uninitialize Aria
Aria::uninit();
Aria::shutdown();
// All done
return(0);
}
```

Compilation and execution: The following line is used for the compilation and execution of the server program in LINUX.

```
g++ -o param -I$ARIA/include - L$ARIA/lib -ldl -pthread
-lAria param.cpp
```

Listing 6.2. Client program

```java
// filename : Params.java
import java.net.*;
import java.io.*;
import java.util.*;
import java.awt.*;
import java.awt.event.*;
import javax.swing.*;

public class Params extends JFrame implements Runnable
{
Socket sc;
DataInputStream dis;
DataOutputStream dos;
String input;
StringTokenizer st;

Label robname,robtype,maxvel,maxrvel,vel,rvel,lvel,
robradius,robdiagonal,batv,posx,posy,head;

public Params()
{
// set the title of the frame here.
setTitle("Robot Parameters");

Container cp = getContentPane();
cp.setLayout(new FlowLayout());

//position y
Box mainbox = new Box(BoxLayout.Y_AXIS);
Box sub[]= new Box[12];

for(int i = 0;i<12;i++)
            sub[i] = new Box(BoxLayout.X_AXIS);

// for robot name
robname=new Label("-");
sub[0].add(new Label("Robot Name:"));
sub[0].add(Box.createGlue());
sub[0].add(robname);
```

```
//for robot type
robtype=new Label("-");
sub[1].add(new Label("Robot Type:"));
sub[1].add(Box.createGlue());
sub[1].add(robtype);

//for robot radius
robradius = new Label("0");
sub[2].add(new Label("Robot Radius:"));
sub[2].add(Box.createGlue());
sub[2].add(robradius);

//for robot diagonal
robdiagonal = new Label("0");
sub[3].add(new Label("Robot Diagonal:"));
sub[3].add(Box.createGlue());
sub[3].add(robdiagonal);

// for max. velocity
maxvel = new Label("0");
sub[4].add(new Label("Max.Translational Velocity:"));
sub[4].add(Box.createGlue());
sub[4].add(maxvel);

// for max rotational velocity
maxrvel = new Label("0");
sub[5].add(new Label("Max.Rotational Velocity:"));
sub[5].add(Box.createGlue());
sub[5].add(maxrvel);

// for velocity
vel = new Label("0");
sub[6].add(new Label("Robot Velocity:"));
sub[6].add(Box.createGlue());
sub[6].add(vel);

// for rotational velocity
rvel = new Label("0");
sub[7].add(new Label("Right wheel velocity:"));
sub[7].add(Box.createGlue());
sub[7].add(rvel);

// for left wheel velocity
lvel = new Label("0");
sub[8].add(new Label("Left wheel velocity:"));
sub[8].add(Box.createGlue());
sub[8].add(lvel);
```

```
//for battery voltage
batv = new Label ("0");
sub[9].add(new Label("Battery Volatage:"));
sub[9].add(Box.createGlue());
sub[9].add(batv);

//for position x
posx = new Label ("0,0");
sub[10].add(new Label("Position(X,Y):"));
sub[10].add(Box.createGlue());
sub[10].add(posx);

// for heading
head = new Label ("0");
sub[11].add(new Label("Heading:"));
sub[11].add(Box.createGlue());
sub[11].add(head);

for(int i= 0;i<12;i++)
mainbox.add(sub[i]);
cp.add(mainbox);
addWindowListener(new WindowAdapter()
{
public void windowClosing(WindowEvent e)
{
System.exit(0);
}
});

// connect the client to the server here.
try
{
int n;
sc=new Socket("192.168.0.9",7775);
dis=new DataInputStream(sc.getInputStream());
dos=new DataOutputStream(sc.getOutputStream());
byte[] b = new byte[1024];

synchronized(dis)
{
n=dis.read(b);
}
input = new String(b,0,n);
System.out.println(n+"Server:"+input);
Thread.sleep(500);
dos.write((new String("recd")).getBytes());
dos.flush();
```

```
// taking the initial robot parameters.
st = new StringTokenizer(input,"|");
robname.setText(st.nextToken());
robtype.setText(st.nextToken());
robradius.setText(st.nextToken());
robdiagonal.setText(st.nextToken());
maxvel.setText(st.nextToken());
maxrvel.setText(st.nextToken());
invalidate();
validate();
}
catch(Exception ex)
{
System.out.println("Its here...its here");
ex.printStackTrace();
}
setResizable(false);
pack();
setVisible(true);
}

// the run time settings are done here.
public void run()
{
int n;
if(sc!=null&&dis!=null)
{
System.out.println("Ready"+sc.toString()+
dis.toString());
try
{
do
{
byte[] b= new byte[1024];

// read the run time robot parameters.
synchronized(dis)
{
n=dis.read(b);
}
input = new String(b,0,n);
System.out.println(n+"Server:"+input);
Thread.sleep(500);
dos.write((new String("recd")).getBytes());
dos.flush();
st = new StringTokenizer(input,"|");
```

```
// set the run time readings in the JAVA Frame

posx.setText(st.nextToken()+","+
st.nextToken());
head.setText(st.nextToken());
vel.setText(st.nextToken());
lvel.setText(st.nextToken());
rvel.setText(st.nextToken());
batv.setText(st.nextToken());
invalidate();
validate();
}while(input!=null);
dis.close();
sc.close();
}
catch(Exception ex)
{
ex.printStackTrace();
}
}
}
public static void main(String arg[])
{
Params p =new Params();
try
{
Thread t = new Thread(p);
t.start();
}
catch(Exception e){}
}
}
```

Program log session

```
Syncing 0
Syncing 1
Syncing 2
Connected to robot.
Name: arcane
Type: Pioneer
Subtype: p2de
Loaded robot parameters from p2de.p
```

```
Opened the server port
Client has connected

Disconnecting from robot.
```

Robot Parameters	_ □ ×
Robot Name:	arcane
Robot Type:	Pioneer
Robot Radius:	250
Robot Diagonal:	120
Max.Translational Velocity:	2200
Max.Rotational Velocity:	500
Robot Velocity:	230
Right wheel velocity:	190
Left wheel velocity:	269
Battery Volatage:	11.9
Position(X,Y):	-712,408
Heading:	115

Fig. 6.3. Parameter program output

6.3 Summary

This chapter highlights the flow chart and the program code to display various parameters of the robot which helps writing programs for real time applications. Fig. 6.3 displays the JAVA Frame that contains the robot parameters, such as the name of the robot ("arcane") and its type ("Pioneer"). The robot has a maximum translational velocity of 2200 mm/sec and the maximum rotational velocity is 500 mm/sec. Similarly, the robot's coordinates are currently (−712 mm, 408 mm) with respect to the starting point. The robot's battery voltage is 11.9 volts, while its heading is 115°. Whenever these parameters change they are reflected in the JAVA frame running on the client machine.

7 Program for BotSpeak

7.1 Introduction

Some mobile robots have loudspeakers through which they announce their decisions, which helps the people in and around to know what the robot is doing or what it is going to do. Sometimes it may announce the instructions to follow which makes the robot applications more interactive and user friendly. This sample client–server program speaks the text entered by the user, which is achieved with the help of the **BotSpeak** library available with the robot. The program running at the client takes the text which the user wants the robot to speak. It then sends it to the server program running in parallel on the robot. The server program then sends this text to the speech engine on the robot, which decodes it and finally sends the modulated electric signal to the loud speaker, so the robot is able to speak what you type at the client. The speech engine follows English grammar. So the text typed at the client is spoken by the robot with the help of the server program running on the robot. The communication between the client and the server is through the common socket used by both.

7.2 Flow Chart and Source Code for BotSpeak Program

The client and server flow charts used for the BotSpeak program are illustrated in Figs. 7.1 and 7.2 and their sample programs written in C++ and Java are shown in Listing 7.1 and Listing 7.2 respectively. The BotSpeak program output with program log session is depicted in Fig. 7.3.

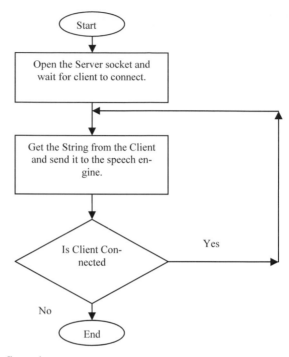

Fig. 7.1. Server program's flow chart

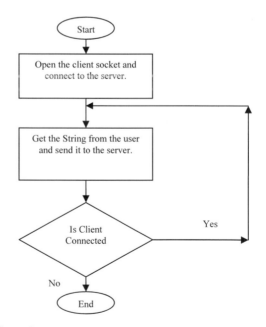

Fig. 7.2. Client program's flow chart

Listing 7.1. Server program

```cpp
// filename : botspeak.cpp
#include<iostream.h>
#include<string.h>
#include"Aria.h"
// for including the Botspeak.h for C
extern "C" {
#include"BotSpeak.h"
}

main()
{
// the Aria client and server socket.
ArSocket server,client;

// the size of the string to be passed.
size_t size;

// the main robot parameter.
ArRobot robot;

// the serial connection is done by this.
ArSerialConnection scon;

// the string for reading the socket.
char str[200];

// for initializing the botspeak machine
bsInit();

// Let the botspeak engine speak the instructions.
bsSpeak("\n Enter any string and I shall speak it ");
// Finish speaking is used for completing the speak in-
struction.
bsFinishSpeaking();

// used to run Aria in a single thread.
Aria::init(Aria::SIGHANDLE_THREAD);

// for opening the socket.
if(scon.open() != 0)
{
printf("\n Could not open the connection.");
exit(1);
}
```

```
robot.setDeviceConnection(&scon);

// wait for the robot to connect.
if(!robot.blockingConnect())
{
printf("\n Could not connect to the robot");
return 1;
}

// open the sever socket.
if(server.open(5001,ArSocket::TCP))
{
printf("\n Opened the Server Socket.");
}
else  // else end the program.
{
printf("\n Unable to open the Server Socket.");
Aria::shutdown();
return 1;
}

// wait for the client to join.
if(server.accept(&client))
printf("\n Connected to Client.");

//  run the robot in asynchronous mode.
robot.runAsync(true);

// loop to collect the string from client and pass it
to the speech engine

while(true)
{
// reading the string from the client.
size = client.read(str,sizeof(str));

// if client is disconnected then exit.
if (size < 0)
break;

if ( size > 0)
{
str[size] = '\0';
printf("\n Client said :: %s %d
%d",str,strlen(str),size );

// finally speaking it here.
```

```
bsSpeak(str);
bsFinishSpeaking();

if(!strcmp(str,"Disconnect"))
{
printf("Disconnecting from the client.");
break;
}
}
}

// closing and uninitialising sockets and others.
client.close();
printf("\n Closing the server");
server.close();
Aria::shutdown();
return 0;
}
```

Compiling and execution: The following text is used for the compilation and execution of the program

```
g++ -o -I$ARIA/include -L$ARIA/lib - I$BOTSPEAK/include
-L$BOTSPEAK/bin // -lAria -ldl -pthread -lbotspeak
botspeak.cpp
```

Listing 7.2. Client Program

```
// filename BotSpeak.java
import java.awt.*;
import javax.swing.*;
import java.net.*;
import java.io.*;
import java.awt.event.*;

public class BOTSPEAK extends JFrame
{
Container container=null;
JTextField text;
String str = "Enter the text you want to hear..";
JLabel label,cLabel;
Socket socket;
InputStream input;
OutputStream output;
```

```java
public BotSpeak(String title)
{
super(title);

container = this.getContentPane();
container.setLayout(null);

// label for displaying the connection status
JLabel cLabel = new JLabel("Not Connected");
cLabel.setBounds(200,280,150,30);
container.add(cLabel);

// label for displaying the text / string
label = new JLabel("",JLabel.CENTER);
label.setBounds(50,150,250,20);
container.add(label);

// text field defined here.
text = new JTextField(str,30);
text.setBounds(50,120,250,20);
text.addActionListener(new TextFieldListener());
container.add(text);

// Clear button for clearing the text-field
JButton clear = new JButton("Clear");
clear.setHorizontalAlignment(SwingConstants.CENTER);
clear.setBounds(140,200,80,25);
clear.addActionListener(new ButtonListener());
container.add(clear);

// open the client socket and connect it to the server
try
{
socket = new Socket("192.168.0.9",5001);
cLabel.setText("Connected to 192.168.0.9");
}
catch(UnknownHostException e)
{
System.out.println(e);
System.exit(ERROR);
}
catch(IOException e)
{
System.out.println(e);
System.exit(ERROR);
}
```

```java
//get the streams for input and output
try
{
output = socket.getOutputStream();
input = socket.getInputStream();
}
catch(IOException e)
{
System.out.println(e);
}

addWindowListener(new WindowEventHandler());
setDefaultCloseOpera-
tion(WindowConstants.DISPOSE_ON_CLOSE)
      setSize(400,350); /// set the size of the frame
      show();
      }

// function for collecting the string from the server.
      private String getString(InputStream in) throws
IOException
{
int c;
int pos =0;
byte buf[]= new byte[1024];
pos = in.read(buf);
if(pos<=0) return null;
String str = new String(buf,0,pos);

return str;
}

//function for writing the string to the socket.
public void writeString(OutputStream o,String s) throws
IOException
{
o.write(s.getBytes());
}

// function that is called when an instance of a class
is being destroyed.
void finalise()
{
try
{
socket.close();
}
```

```
catch(IOException e)
{
System.out.println(e);
}
}

// class used for implementing the window features
(window closing)
class WindowEventHandler extends WindowAdapter
{
public void windowClosing(WindowEvent e)
{
try
{
writeString(output,"Disconnect");
}
catch(IOException exc){}
System.exit(0);
}
}

// class used to implement action listener to the text
field.
class TextFieldListener implements ActionListener
{
public void actionPerformed(ActionEvent e)
{
label.setText(e.getActionCommand());
try
{
writeString(output,e.getActionCommand());
}
catch( IOException exp) {}
}
}
// class for implementing the action listener for the
buttons.
class ButtonListener implements ActionListener
{
public void actionPerformed(ActionEvent e)
{
```

```
text.setText("");
label.setText("");
text.requestFocus();
}
}
// the main function.
public static void main(String[] args)
{
BotSpeak frame = new BotSpeak("BotSpeak");
}
}
```

Program log session

```
Syncing 0
Attempting to close previous connection.
Syncing 0
Syncing 1
Syncing 2
Connected to robot.
Name: arcane
Type: Pioneer
Subtype: p2de
Loaded robot parameters from p2de.p

Opened the Server Socket.
Connected to Client.

Client said :: Hello! Welcome to the world of Robotics.
40 40
Client said :: India is my motherland 22 22
Client said :: Connected to 192.168.0.9 24 24
Client said :: Disconnect 10 10

Disconnecting from the client.
Closing the server
Disconnecting from robot.
```

Fig. 7.3. Output from the BotSpeak program

7.3 Summary

The chapter highlights the usefulness of the BotSpeak program. Fig. 7.3 shows the JAVA Frame consisting of a text box and a "Clear" button. The robot pronounces the text written in the text field. The IP address of the server, i.e. 192.168.0.9 is displayed at the bottom right-hand side of the display window.

8 Gripper Control Program

8.1 Introduction

Modern robot applications include tasks such as collecting garbage, holding objects etc. with the help of gripper commands of the robot. Here a client–server program has been developed for the use of the gripper control. The client program takes the input request, such as a gripper, and lift operation, like "gripper open" or "lift up" etc. Then the appropriate command in the form of a string is passed on to the server running on the robot, which ultimately passes the command to the microcontroller, which performs the requested operation. The client can perform any of the gripper operations such as gripper open, gripper close, gripper stop, lift up, lift down, lift stop, ready, stop, using standard functions defined on the robot server.

8.2 Flow Chart and Source Code for Gripper Control Program

The client and server flow charts used for the gripper control program are illustrated in the Figs. 8.1 and 8.2 and their sample programs written in C++ and JAVA are shown in Listing 8.1 and Listing 8.2 respectively. The gripper control program output with program log session is depicted in Fig. 8.3.

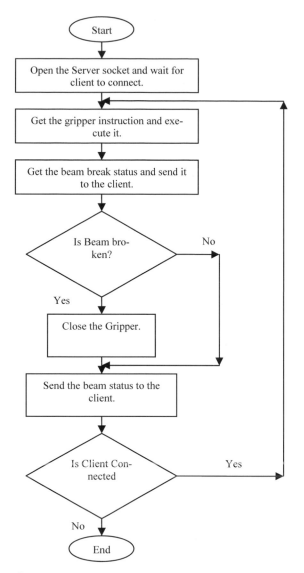

Fig. 8.1. Flow chart of server program

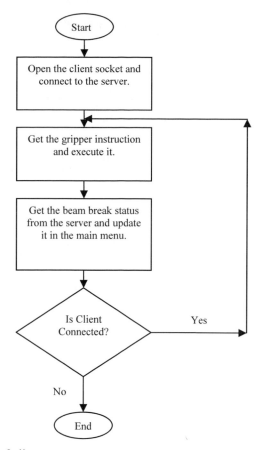

Fig. 8.2. Flow chart of client program

Listing 8.1. Server source code for gripper control

```
// filename : gripper.cpp

#include"Aria.h"
#include<stdlib.h>
#include<signal.h>

ArSocket server,client;

// for ctrl+c used to terminate the program in between
void shutdown(int signum)
```

```
{
printf("\n Closing the connection");
client.close();
server.close();
Aria::shutdown();
exit(0);
}

main()
{
// used for adding ctrl+c handler
struct sigaction sa;
memset(&sa,0,sizeof(sa));
sa.sa_handler = &shutdown;
sigaction(SIGINT,&sa,NULL);

// size of the received string
size_t size;

// the main object i.e. the robot
ArRobot robot;

// sconn for serial connection
ArSerialConnection scon;

// grip used for including gripper in the program
ArGripper grip(&robot);

char str[100];

// running the robot in a single thread.
Aria::init(Aria::SIGHANDLE_THREAD);

// opening the robot connection.
if(scon.open() != 0)
{
printf("\n Could not open the connection.");
exit(1);
}

robot.setDeviceConnection(&scon);

// connecting the robot
if(!robot.blockingConnect())
{
printf("\n Could not connect to the robot");
}
```

```
// opening the server socket.
if(server.open(5011,ArSocket::TCP))
{
printf("\n Opened the Server Socket.");
}
else
{
printf("\n Unable to open the Server Socket.");
Aria::shutdown();
return 1;
}

// waiting for the client to join.
if(server.accept(&client))
{
printf("\n Connected to Client.");
}

// running the robot in asynchronous mode
robot.runAsync(true);

// activate the robots motors.
robot.enableMotors();

while(true)
{
// read the instruction from the client
size = client.read(str,sizeof(str));

// if client disconnected then break the loop
if (size < 0)
break;

if ( size > 0)
{
str[size] = '\0';
robot.lock();

// if instruction = Open then open the gripper pads.
if (!strcmp(str,"Open"))
{
grip.gripOpen();
ArUtil::sleep(50);
}

// if instruction = Close then close the gripper pads.
if (!strcmp(str,"Close"))
```

```
{
grip.gripClose();
ArUtil::sleep(50);
}

// if instruction = "Up" then move the gripper up.
if (!strcmp(str,"Up"))
{
grip.liftUp();
ArUtil::sleep(50);
}

// if instruction = "Down" then move the gripper up
if (!strcmp(str,"Down"))
{
grip.liftDown();
ArUtil::sleep(50);
}

// if instruction = "L Stop" Stop the lift
if (!strcmp(str,"LStop"))
{
grip.liftStop();
ArUtil::sleep(50);
}

// if instruction = "GStop" stop the gripper
if(!strcmp(str,"GStop"))
{
grip.gripStop();
ArUtil::sleep(50);
}

// if instruction = Store then place the gripper in
store pos.
if(!strcmp(str,"Store"))
{
grip.gripperStore();
ArUtil::sleep(50);
}

// if instruction = Ready then place the gripper in
ready pos.
if (!strcmp(str,"Ready"))
{
grip.gripperDeploy();
ArUtil::sleep(50);
}
```

```
// if instruction = "Disconnect" then close sockets and
exit
if(!strcmp(str,"Disconnect"))
{
printf("Disconnecting from the client.");
client.close();
server.close();
Aria::shutdown();
return 0;
}
robot.unlock();
}

// get the beam break status.
str[0] = '\0';
robot.lock();
sprintf(str,"%d",grip.getBreakBeamState());

// if beam is broken then close the gripper
if(grip.getBreakBeamState())
{
grip.gripClose();
ArUtil::sleep(50);
}
robot.unlock();
//send the beam -break state.
client.write(str,strlen(str));
}
printf("\n Closing the server");
client.close();
server.close();
Aria::shutdown();
return 0;
}
```

Compilation and execution: The following text command is used for

compilation and execution.

```
g++ -o gripper -I$ARIA/include -L$ARIA/lib -ldl -
pthread -lAria gripper.cpp
```

Listing 8.2. Client source code for gripper control

```java
//filename : Gripper.java
import java.awt.*;
import javax.swing.*;
import java.io.*;
import java.net.*;
import java.awt.event.*;
import java.text.*;
import java.util.*;

public class Gripper extends JFrame implements Ac-
tionListener
{
Container container=null;
boolean flag = true;
Socket socket;
InputStream input;
OutputStream output;
String string1 = "";

// the labels for various buttons
String str[] = {
"Gripper Open","Gripper Close","Lift Up","Lift Down",
"Grip Stop","Lift Stop","Store","Ready"
          };

JLabel cLabel;
char mnemonics[] = {'O','C','U','D','G','L','S','R'};
int bWidth = 120;
int bHeight = 25;

//location of various buttons
int loc [][] = {
{70,160},{210,160},{140,130},{140,190},
{10,240},{270,240},{10,210},{270,210}
};

public Gripper(String title)
{
super(title);

container = this.getContentPane();
container.setLayout(null);

//adding all the buttons.
for ( int i = 0 ; i < str.length ; i++ )
```

```
{
JButton b = new JButton(str[i]);
b.setBounds(loc[i][0],loc[i][1],bWidth,bHeight);
b.setMnemonic(mnemonics[i]);
container.add(b);
b.addActionListener(this);
}

// label for showing connection status.
JLabel cLabel = new JLabel("Not Connected");
cLabel.setBounds(200,280,150,30);
container.add(cLabel);

// new label where the beams will be dispalyed
Lbl lbl = new Lbl();
lbl.setBounds(0,0,399,130);
container.add(lbl);

// open the client socket and connect to the server.
try
{
socket = new Socket("192.168.0.9",5011);
cLabel.setText("Connected to 192.168.0.9");
}
catch(UnknownHostException e)
{
System.out.println(e);
System.exit(ERROR);
}
catch(IOException e)
{
System.out.println(e);
System.exit(ERROR);
}

// get the input and the output streams
try
{
output = socket.getOutputStream();
input = socket.getInputStream();
}
catch(IOException e)
{
System.out.println(e);
}
```

```
addWindowListener(new WindowEventHandler());
 setDefaultCloseOpera-
tion(WindowConstants.DISPOSE_ON_CLOSE)
setSize(400,350); //set the size of the frame
show();

String string="0";//for no beam break state.
try
{
while(flag)
{
// for beam break state.

if(string.equals("1")||string.equals("2")||string.equal
s("3"))
{
writeString(output,"Close");
string = "0";
}
else
{
// do nothing
if (string1.equals(""))
{
writeString(output,"Hello");
}
else
{
writeString(output,string1);
System.out.println(string1);
string1 = "";
}
}
string = getString(input);
lbl.setString(string);
lbl.repaint();
}
}
catch(IOException e){}
}

// function for performing the action when buttons are
pressed.
public void actionPerformed(ActionEvent ae)
{
String s = ae.getActionCommand();
```

```
if (s.equals(str[0]))    string1 = "Open";
if (s.equals(str[1]))    string1 = "Close";
if (s.equals(str[2]))    string1 = "Up";
if (s.equals(str[3]))    string1 = "Down";
if (s.equals(str[4]))    string1 = "GStop";
if (s.equals(str[5]))    string1 = "LStop";
if (s.equals(str[6]))    string1 = "Store";
if (s.equals(str[7]))    string1 = "Ready";
}

// destructor function
void finalise()
{
try
{
socket.close();
}
catch(IOException e)
{
System.out.println(e);
}
}

// function to get the string from the server.
private String getString(InputStream in) throws IOEx-
ception
{
int c;
int pos =0;
byte buf[]= new byte[1024];

pos = in.read(buf);
if(pos<=0) return null;
String str = new String(buf,0,pos);
return str;
}

// function to write the string to the socket.
public void writeString(OutputStream o,String s) throws
IOException
{
o.write(s.getBytes());
}

// class to implement window functions.
class WindowEventHandler extends WindowAdapter
{
```

```
public void windowClosing(WindowEvent e)
{
flag = false;
try
{
writeString(output,"Disconnect");
}
catch(IOException exc){}
System.exit(0);
}
}

// finally the main
public static void main(String args[])
{
Gripper frame = new Gripper("Gripper");
}

// class for displaying the beams within a Jlabel.
class Lbl extends JLabel
{
public String s="0";
public void setString(String s1)
{
s = s1;
}
public void paint(Graphics g)
{
g.setColor(Color.black);
g.drawString("Inner",10,60);
g.drawString("Inner",360,60);
g.drawString("Outer",10,110);
g.drawString("Outer",360,110);
g.setColor(Color.yellow);

if (s.equals("0"))
{
g.fillRect(50,50,300,10); // inner beam
g.fillRect(50,100,300,10);//outer beam
}
if(s.equals("1"))
{
g.fillRect(50,50,140,10); // inner beam broken
g.fillRect(210,50,140,10);//inner beam broken
g.fillRect(50,100,300,10);//outer beam
}
if(s.equals("2"))
```

```
{
g.fillRect(50,50,300,10); // inner beam
g.fillRect(50,100,140,10); //outer beam broken
g.fillRect(210,100,140,10); //outer beam broken
}
if(s.equals("3"))
{
g.fillRect(50,50,140,10); // inner beam broken
g.fillRect(210,50,140,10); // inner beam broken
g.fillRect(50,100,140,10); //outer beam broken
g.fillRect(210,100,140,10); // outer beam broken
}
}
}
}
```

Program log session

```
Syncing 0
Syncing 1
Syncing 2
Connected to robot.
Name: arcane
Type: Pioneer
Subtype: p2de
Loaded robot parameters from p2de.p

Opened the Server Socket.
Connected to Client.Gripper:  querried, using General
IO.
Disconnecting from the client.
Aria: Received signal 'SIGINT'. Shutting down.
Disconnecting from robot.

Closing the connection
```

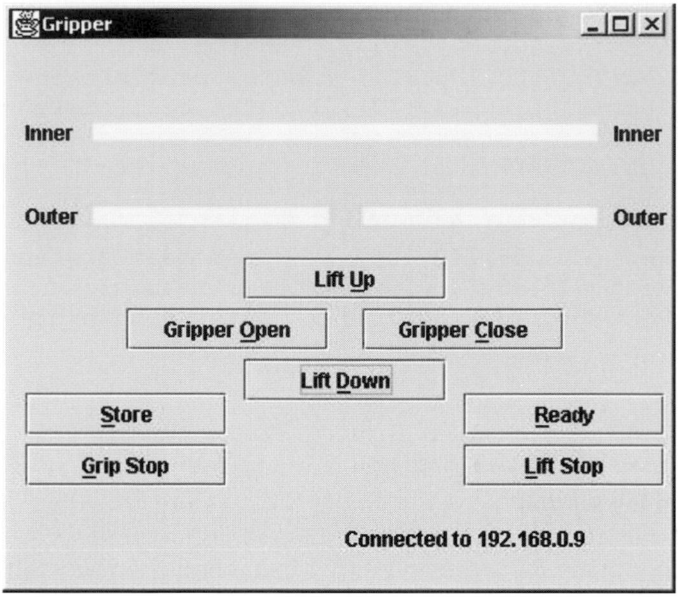

Fig. 8.3. Output of the gripper control program

8.3 Summary

In this chapter the concept of gripper control is explained with the related programs, which is useful in real-time applications. There are two beams in the gripper, one at the outer end of the gripper while the other is at the opposite extreme, i.e. inner end. When they break it means that there is some obstacle between the grippers, and the gripper takes action to close it. If any beam is broken it is shown as a broken beam in the output JAVA Frame as in Fig. 8.3. As shown there are eight buttons. When any one of them is clicked then the client receives the user request. This information is forwarded to the server where the corresponding subroutine program runs. When "Grip Open" is pressed, it opens the gripper. If "Grip Close" is pressed then the gripper will be closed. If "Grip Stop" is pressed, then the moving gripper will be stopped and the same process is carried out for lifting the gripper. The function of "Ready" keeps the gripper in the ready position i.e. ready to grab any object laying in front it. The "Store" key keeps the gripper in store position. The break in the outer beam shown in the Fig. 8.3 is due to the presence of an obstacle between the gripper arms.

9 Program for Sonar Reading Display

9.1 Introduction

Sonar (Sound Navigation and Ranging) is an integral part of any mobile robot; it helps it to know how far the obstacles are located in different directions. In Pioneer 2DE, there are 16 sonar detectors, which are placed uniformly around the robot. With the help of these sonars the robot is able to estimate the distance of the obstacles in every direction and by doing so the robot can avoid obstacles while wandering or navigating.

A sample client–server program has been developed for this purpose, where the server program runs on the robot and client program on the client. The server program gets the sonar reading from each of the sonar sensors and sends these readings one after another to the client. After receiving these sonar readings the client displays them within a JAVA frame graphically, so that all the sonar readings can be simultaneously observed on the client. This process continues, and changes in the sonar readings are reflected within the frame shown on the client.

9.2 Flow Chart and Source Code for Sonar Reading Display on Client

The client and server flow charts used for the sonar display program are illustrated in Figs. 9.1 and 9.2 and their sample programs written in C++ and Java are shown in Listing 9.1 and Listing 9.2 respectively. The program output with program log session is depicted in Fig. 9.3.

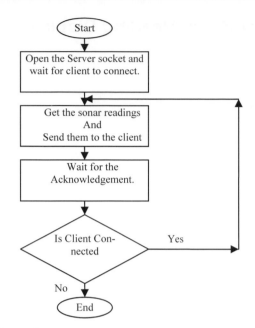

Fig. 9.1. Flow chart of server program

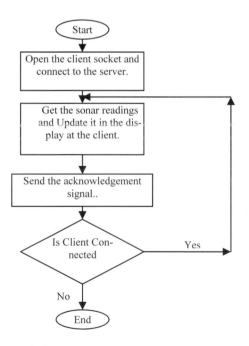

Fig. 9.2. Flow chart of client program

Listing 9.1. Server source code for sonar reading display

```cpp
// filename : sonar.cpp
#include"Aria.h"
#include<stdlib.h>
#include<string.h>
#include<signal.h>

ArSocket server,client;

// for ctrl+c
void shutdown(int signum)
{
printf("\n Closing the connection");
client.close();
server.close();
Aria::shutdown();
exit(0);
}

main()
{
// for ctrl + C handler
struct sigaction sa;
memset(&sa,0,sizeof(sa));
sa.sa_handler = &shutdown;
sigaction(SIGINT,&sa,NULL);

size_t size;

// scon for serial connection
ArSerialConnection scon;

// the robot object
ArRobot robot;

// sonar is used for getting sonar readings.
ArSonarDevice sonar;

// for storing the sonar readings
int range[16];

// for sending the sonar readings through str.
char str[100];

// Running in a single thread
Aria::init(Aria::SIGHANDLE_THREAD);
```

```
// opening the robot connection
if(scon.open() != 0)
{
printf("\n Could not open the connection.");
exit(1);
}
else
{
printf("\n Connected to robot through serial port.");
robot.addRangeDevice(&sonar);
robot.setDeviceConnection(&scon);
}

// connecting the robot.
if (!robot.blockingConnect())
{
printf("\n Could not connect to robot");
Aria::shutdown();
exit(1);
}

//opening the server socket.
if(server.open(5015,ArSocket::TCP))
printf("\n Opened the Server Socket.");
else
{
printf("\n Unable to open the Server Socket.");
return 1;
}

// wait for client to join
if(server.accept(&client))
printf("\n Connected to Client.");

// run robot in asynchronous mode.
robot.runAsync(true);

// sonar readings are send here.
while (true)
{
for ( int i = 0 ; i < 16 ; i ++)
{
// get the ready / ack signal
size = client.read(str,sizeof(str));
str[size] = '\0';
```

```
// if Disconnect then break the loop and exit.
if(!strcmp(str,"Disconnect"))
{
client.close();
server.close();
Aria::shutdown();
return 0;
}

str[0] = '\0';

// getting the sonar reading
range[i] = robot.getSonarRange(i);
sprintf(str,"%d",range[i]);

// send it to the client
client.write(str,strlen(str));
}
}

client.close();
server.close();
Aria::shutdown();
return 0;
}
```

Compilation and execution: The following text command is used for compilation and execution.

```
g++ -o sonar - I$ARIA/include -L$ARIA/lib  -ldl -
pthread -lAria sonar.cpp
```

Listing 9.2. Client source code for sonar reading display

```
// filename : Sonar.java
import java.awt.*;
import javax.swing.*;
import java.io.*;
import java.net.*;
import java.awt.event.*;
import java.text.*;
import java.lang.*;
import java.util.*;
public class Sonar extends JFrame
```

```
{
Container container=null;
boolean flag =true;
Socket socket;
InputStream input; OutputStream output;
JLbl lbl;
double s_r[] = new double[16];

public Sonar(String title)
{
super(title);
container = this.getContentPane();
container.setLayout(null);

// label for connection status
JLabel cLabel = new JLabel("Not Connected");
cLabel.setBounds(180,310,150,30);
container.add(cLabel);
lbl = new JLbl();
lbl.setBounds(0,0,350,350);
container.add(lbl);

// connecting to the server
try
{
socket = new Socket("192.168.0.9",5015);
cLabel.setText("Connected to 192.168.0.9");
}
catch(UnknownHostException e)
{
System.out.println(e);
System.exit(ERROR);
}
catch(IOException e)
{
System.out.println(e);
System.exit(ERROR);
}

// opening the input and output streams of the socket
try
{
output = socket.getOutputStream();
input = socket.getInputStream();
}
catch(IOException e)
```

```
{
System.out.println(e);
}

addWindowListener(new WindowEventHandler());
setDefaultCloseOpera-
tion(WindowConstants.DISPOSE_ON_CLOSE)
setSize(330,380);
show();

String str; // for taking the input

try
{
while(true)
{
for (int i = 0 ; i < 16 ; i++ )
{
if (flag)
{// sending the ack. Signal here
writeString(output,"Hello");
str = getString(input);
s_r[i] = (double)Integer.parseInt(str);
}
}
lbl.repaint();
}
}
catch(IOException e){}
}

// the destructor
void finalise()
{
try
{
socket.close();
}
catch(IOException e)
{
System.out.println(e);
}
}

// the function of obtaining input string from the
socket
private String getString(InputStream in) throws IOEx-
ception
```

```
{
int c, pos =0;
byte buf[]= new byte[1024];
pos = in.read(buf);
if(pos<=0) return null;
String str = new String(buf,0,pos);
return str;
}

// function for writing data to the socket.
public void writeString(OutputStream o,String s) throws
IOException
{
o.write(s.getBytes());
}

// class for implementing window features.
class WindowEventHandler extends WindowAdapter
{
// while closing the clinet window send "DISCONNECT"
signal
public void windowClosing(WindowEvent e)
{
flag = false;
try
{
writeString(output,"Disconnect");
}
catch(IOException exc){}
System.exit(0);
}
}

// the main function
public static void main(String args[])
{
Sonar frame = new Sonar("Sonar");
}

// the label where the graphics i.e. sonar readings
will be displayed
class JLbl extends JLabel
{
public int width = 350,height= 350;
public int cx = 155,cy = 155,d = 14;
int ang[] = {160,140,120,100,80,60,40,20,
       340,320,300,280,260,240,220,200};
```

```
double s_max;        double x,y;

public JLbl()
{
s_max = 5000.0;
for ( int i = 0 ; i < 16 ; i ++ )
s_r[i] = s_max;
}

public void paint(Graphics g)
{
g.setColor(Color.white);
g.fillOval(10,10,290,290); // for drawing the range of
sonars.

// the outer boundary of sonars.
g.setColor(Color.black);
g.drawOval(10,10,290,290);

// the robot at the center
g.setColor(Color.black);
g.drawOval(cx-d/2,cx-d/2,d,d);
g.drawLine(cx,cy,cx,cy-d/2);

// for 16 sonars
for ( int i = 0 ; i < 16 ; i++ )
{
if ( s_r[i] > s_max )
s_r[i] = s_max;

// calculating the x and y coordinates
x = 140*Math.cos(Math.toRadians(ang[i]))*s_r[i]/s_max;
y = 140*Math.sin(Math.toRadians(ang[i]))*s_r[i]/s_max;

if (x > 0) x += d/2;
else   x -= d/2;
if ( y > 0) y += d/2;
else   y -= d/2;

// draw a cross at the location (x,y)
cross(g);
}
}

// for drawing the cross
public void cross(Graphics g)
{
```

```
int x1,y1;
if (x > 0)   x1 = (int)(x+0.5);
else         x1 =(int)(x-0.5);
if (y > 0)   y1 = (int)(y+0.5);
else         y1 =(int)(y-0.5);
g.setColor(Color.blue);
g.drawLine(cx+x1-2,cy-y1-2,cx+x1+2,cy-y1+2);
g.drawLine(cx+x1+2,cy-y1-2,cx+x1-2,cy-y1+2);
}
//function for getting the width of the label
public int getIconWidth(){return width;}
//function for getting the height of the label
public int getIconHeight(){return height;}
}
}
```

Program log session

```
Connected to robot through serial port.Syncing 0
Syncing 1
Syncing 2
Connected to robot.
Name: arcane
Type: Pioneer
Subtype: p2de
Loaded robot parameters from p2de.p
Opened the Server Socket.
Connected to Client.
Disconnecting from robot.
```

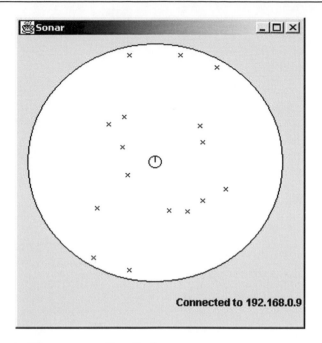

Fig. 9.3. Output of the sonar reading display program

9.3 Summary

In this chapter the program for the sonar reading display has been discussed. It can be seen from Fig. 9.3, that there are 16 crosses in total in the figure. Each of these crosses represents a sonar reading. The robot is shown at the center of the frame. There is an outer circle, which shows the upper limit of sonar readings. The sonar readings shown in the figure represent the actual sonar positions of the robot. The label at the bottom right side shows that the client is connected to the server. This program will be helpful for the development of various application programs for mobile robots to avoid obstacle collision. We will discuss such a program in the next chapter, i.e. wandering with obstacle avoidance by using the sonar readings.

10 Program for Wandering Within the Workspace

10.1 Introduction

One of the real-time applications of a robot is to make the robot wander freely without colliding with obstacles. The robot is made to wander freely around with the help of actions. Here actions are synchronous tasks which the robot does while executing the program. Actions are added to make the robot move in "Wander" mode. There are actions like action `Avoid-FrontNear`, action `AvoidFrontFar`, action `ConstantVelocity`, action `Avoid Bumpers` etc., which are available in ARIA. Simply by adding these actions, the robot runs in wander mode. This is so, because the robot simultaneously checks for front avoidance distance, side distance, the bumpers and the other actions, which are being added. So, according to the actions added the robot could freely move without bumping into anything. Actions are different from Direct motions, which follow instructions asynchronously. If there is a move command the robot will try to move even if there is an obstacle, while in actions the robot decides itself about the motion depending on the run-time conditions.

10.2 Algorithm and Source Code for Wandering Within the Workspace

The server algorithm implements the actions on the robot when the signal "Start" comes from the client. The client algorithm takes the instructions from the user regarding starting/stopping the wander mode. The client and server flow charts used for wandering in the workspace are illustrated in the Figs. 10.1 and 10.2 and their sample programs written in C++ and Java are shown in Listing 10.1 and Listing 10.2 respectively. The program output with program log session is depicted in Fig. 10.3.

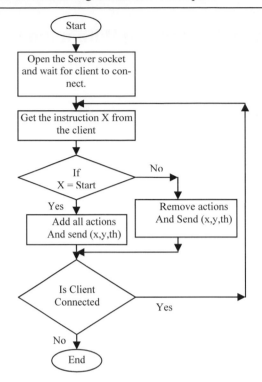

Fig. 10.1. Server flow chart for the wander program

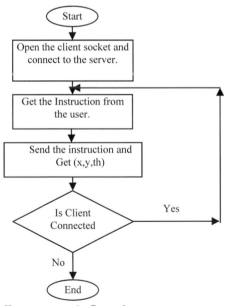

Fig. 10.2. Wander client program's flow chart

Listing 10.1. Server source code for wander program

```cpp
// filename : Wander.cpp
#include "Aria.h"
#include <math.h>
#include <string.h>
#include<signal.h>
#include<stdlib.h>
ArSocket server,client;

// for ctrl+c
void shutdown(int signum)
{
printf("\n Closing the connection");
client.close();server.close();
Aria::shutdown();
exit(0);
}
// the main
main()
{
// the ctrl+c handler
struct sigaction sa;
memset(&sa,0,sizeof(sa));
sa.sa_handler = &shutdown;
sigaction(SIGINT,&sa,NULL);

// for serial connection
ArSerialConnection scon;

// for sonar readings sonar device needs to be added
ArSonarDevice sonar;

// the robot object
ArRobot robot;

// the actions
ArActionStallRecover recover;
ArActionBumpers bumpers;
ArActionAvoidFront avoidFrontNear("Avoid Front Near",
275, 0);
ArActionAvoidFront avoidFrontFar;
ArActionConstantVelocity constantVelocity("Constant Ve-
locity", 250);

// run the program in single thread
Aria::init(Aria::SIGHANDLE_THREAD);
```

```
// open the robot connection
if(scon.open() != 0)
{
printf("\n Could not open the connection.");
exit(1);
}
robot.setDeviceConnection(&scon);
robot.addRangeDevice(&sonar);

// connect the robot
if (!robot.blockingConnect())
{
printf("Could not connect to robot... exiting\n");
Aria::shutdown();
return 1;
}
// open the server socket.
if(server.open(4444,ArSocket::TCP))
printf("\n Opened the Server Socket.");
else  printf("\n Unable to open the Server Socket.");

// wait for the client to join
if(server.accept(&client))
printf("\n Connected to Client.");

// enable the robot motors.
robot.comInt(ArCommands::ENABLE, 1);
robot.comInt(ArCommands::SOUNDTOG, 0);

// run the robot in asynchronous mode.
robot.runAsync(true);

while(true)
{
char str[20];

// get the instruction fomr the client
int size = client.read(str,sizeof(str));

str[size]='\0';

// if that is start then add actions
if(!strcmp(str,"START"))
{
robot.lock();
robot.clearDirectMotion();
```

```
robot.addAction(&recover, 100);
robot.addAction(&bumpers, 75);
robot.addAction(&avoidFrontNear, 50);
robot.addAction(&avoidFrontFar, 49);
robot.addAction(&constantVelocity, 25);
robot.unlock();
}

// else remove actions
else if(!strcmp(str,"STOP"))
{
robot.lock();
robot.remAction(&recover);
robot.remAction(&bumpers);
robot.remAction(&avoidFrontNear);
robot.remAction(&avoidFrontFar);
robot.remAction(&constantVelocity);
robot.stop();
robot.unlock();
}
// if exit then exit
else if(!strcmp(str,"EXIT"))
{
robot.remAction(&recover);
robot.remAction(&bumpers);
robot.remAction(&avoidFrontNear);
robot.remAction(&avoidFrontFar);
robot.remAction(&constantVelocity);
client.close();server.close();Aria::shutdown();
break;
}

sprintf(str,"%d|%d|%d|",(int)robot.getX(),
(int)robot.getY(),(int)robot.getTh());

//send the (x,y,th);
client.write(str,strlen(str));
}
return 0;
}
```

Compilation and execution: The following text command is used for compilation and execution of the wander server program.

```
g++ -o -I$ARIA/include  -L$ARIA/lib -ldl -pthread -
lAria wander.cpp
```

Listing 10.2. Client source code for wander program

```java
filename : Wander.java
import java.awt.*;
import javax.swing.*;
import java.io.*;
import java.net.*;
import java.awt.event.*;
import java.text.*;
import java.lang.*;
import java.util.*;

public class Wander extends JFrame
{
Container container=null;
Socket socket;
InputStream input;
OutputStream output;
JLabel x,y,th;
double x1=0.0,y1=0.0,th1=0.0;
JLbl lbl;

public Wander(String title)
{
super(title);
container = this.getContentPane();
container.setLayout(null);

// label for connection status
JLabel cLabel = new JLabel("Not Connected");
cLabel.setBounds(200,280,150,30);
container.add(cLabel);

// labels for (X,Y,theta)
x = new JLabel("X :: 0.0");
y = new JLabel("Y :: 0.0");
th = new JLabel("Th :: 0.0");
x.setBounds(10,250,100,15);
y.setBounds(10,265,100,15);
th.setBounds(10,280,50,15);
container.add(x);
container.add(y);
container.add(th);

// the label where the simulator is displayed
lbl = new JLbl();
lbl.setBounds(0,0,399,170);
container.add(lbl);
```

```
// the socket is opened here.
try
{
socket = new Socket("192.168.0.9",5007);
cLabel.setText("Connected to 192.168.0.9");
}
catch(UnknownHostException e)
{
System.out.println(e);
System.exit(ERROR);
}
catch(IOException e)
{
System.out.println(e);
System.exit(ERROR);
}
// get the input and output streams.
try
{
output = socket.getOutputStream();
input = socket.getInputStream();
}
catch(IOException e)
{
System.out.println(e);
}
// add the window features.
addWindowListener(new WindowEventHandler());
setDefaultCloseOpera-
tion(WindowConstants.DISPOSE_ON_CLOSE)
setSize(400,350);
show();

// three variables for (x,y,th)
double z[] = new double[3];
String str;

try
{
while(true)
{
// getting the (x,y.th) here
for ( int i = 0 ; i < 3 ; i++  )
{
writeString(output,"Hello");
```

```
str = getString(input);
z[i] = (double)Integer.parseInt(str);
}

// using an appropriate scale
x1 = z[0]/100;
y1 = z[1]/100;
th1 = z[2]/100;

x.setText("X :: "+x1);
y.setText("Y :: "+y1);
th.setText("Th :: "+th1);
lbl.setValues(x1,y1,th1);
// when values change repaint the label.
lbl.repaint();
}
}
catch(IOException err){}
}
// destructor
void finalise()
{
try
{
socket.close();
}
catch(IOException e)
{
System.out.println(e);
}
}
// function for getting the string/data from socket
private String getString(InputStream in) throws IOEx-
ception
{
int c; int pos =0;
byte buf[]= new byte[1024];
pos = in.read(buf);
if(pos<=0) return null;
String str = new String(buf,0,pos);
return str;
}

// function to write data to the socket.
public void writeString(OutputStream o,String s) throws
IOException
{
```

```
o.write(s.getBytes());
}

// class to implement the window functions.
class WindowEventHandler extends WindowAdapter
{
public void windowClosing(WindowEvent e)
{
try
{
writeString(output,"Disconnect");
}
catch(IOException exc){}
System.exit(0);
}
}

// the main function.
public static void main(String args[])
{
Wander frame = new Wander("Wander");
}

// the lablel where the simulator is displayed
class JLbl extends JLabel
{
public int width = 380,height= 200;
public int x=197,y=97,d=14,th=0;
public int cx=x,cy=y;

public void TLbl()
{
}

// the simulator is displayed using paint(g)
public void paint(Graphics g)
{
// the outer boundary of the simulator.
g.setColor(Color.white);
g.fillRect(10,10,370,170);
g.setColor(Color.black);
g.drawRect(10,10,370,159);

// the robot
g.drawOval(cx-d/2,cy-d/2,d,d);
int x1 = (int)(d*Math.cos(Math.toRadians(th))/2);
int y1 = (int)(d*Math.sin(Math.toRadians(th))/2);
```

```
g.drawLine(cx,cy,cx+x1,cy-y1);
}
// function to return width.
public int getIconWidth(){return width;}

// function to return height.
public int getIconHeight(){return height;}

//function to set values.
public void setValues(double x1,double y1,double th1)
{
x1 %= 8;y1 %= 3;
cx = x + (int)(x1*20);
cy = y - (int)(y1*20);
th = (int)th1;
}
}
}
```

Program log session

```
Syncing 0
Syncing 1
Syncing 2
Connected to robot.
Name: arcane
Type: Pioneer
Subtype: p2de
Loaded robot parameters from p2de.p

Opened the Server Socket.
Connected to Client.
Disconnecting from robot.
```

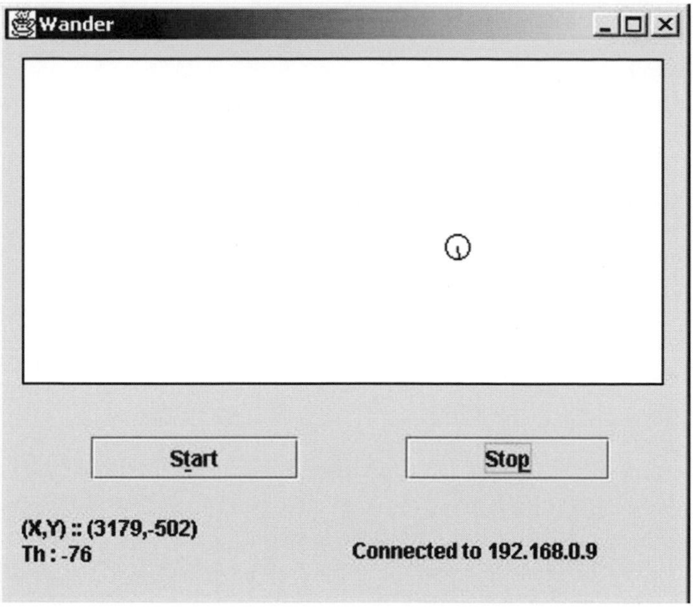

Fig. 10.3. Output of the wander program

10.3 Summary

As shown in Fig. 10.3, the client interface has two buttons, i.e. "Start" and "Stop". When one clicks the "Start" button then the robot starts running in "Wander" mode and when the "stop" button is clicked the robot stops wandering. The current position and heading (Th) in the figure are indicated in the bottom left corner of the JAVA frame and at the right-hand corner there is a label showing that the robot is connected to the client.

11 Program for Tele-operation

11.1 Introduction

Tele-operation is used for operating the robot from a distance. A common example is a pick and place application, where the robot is controlled by the master (user) from a control room. Here a sample client–server program is developed, which controls the robot from the client. In the client the motion control is done by using the keys or the buttons in the menu, which is displayed at the client in a simulator. The simulator has the current robot location and the user at the client is able to know exactly where the robot is, along with its heading.

11.2 Algorithm and Source Code for Tele-operation

The server algorithm is meant for collecting the commands regarding motion from the client and executing them. The client algorithm gives the instructions to get these commands from the user. The client and server flow charts used for the program for tele-operation are illustrated in Figs. 11.1 and 11.2 and their sample programs written in C++ and Java are shown in Listings 11.1 and 11.2 respectively. The program output with program log session is depicted in Fig. 11.3.

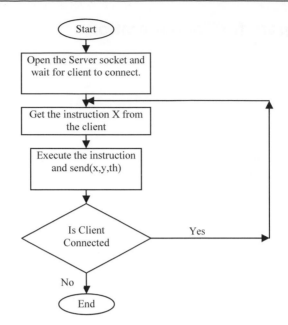

Fig. 11.1. Server flow chart for the program for tele-operation

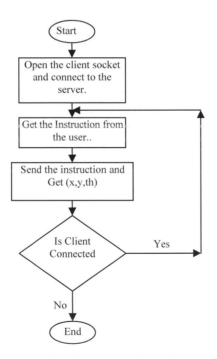

Fig. 11.2. Client flow chart for the program for tele-operation

Listing 11.1. Server source code for tele-operation program

```cpp
// filename : teleop.cpp
#include"Aria.h"
main()
{
// for socket
ArSocket server,client;

// size of read data
size_t size;

// the robot instance
ArRobot robot;

// the serial connection
ArSerialConnection scon;

char str[100];

// for forward step / Back move / Turn
int fStep = 100;int bStep = 100;
int lHeading = 10;int rHeading = 10;
int head=0;

// run program in single thread
Aria::init(Aria::SIGHANDLE_THREAD);

// open the connection
if(scon.open() != 0)
{
printf("\n Could not open the connection.");
exit(1);
}

// set the serial connection
robot.setDeviceConnection(&scon);

// connect the robot
if(!robot.blockingConnect())
printf("\n Could not connect to the robot");

// the server socket
if(server.open(5000,ArSocket::TCP))
printf("\n Opened the Server Socket.");
else
```

```
{
printf("\n Unable to open the Server Socket.");
Aria::shutdown();
return 1;
}

// wait for the client to join
if(server.accept(&client))
printf("\n Connected to Client.");

robot.runAsync(true);

// enable the motors
robot.enableMotors();

while(true)
{
// get the instruction
size = client.read(str,sizeof(str));

// if client disconnected then break loop
if (size < 0)       break;

if ( size > 0)
{
str[size] = '\0';
printf("\n Client said :: %s %d
%d",str,strlen(str),size );

// for forward instruction
if (!strcmp(str,"Forward"))
{
robot.lock();
robot.move(fStep);
robot.unlock();
while(true)
{
robot.lock();
if(robot.isMoveDone())
{
robot.unlock();
break;
}
robot.unlock();
ArUtil::sleep(50);
}
printf("\n %lf %lf %lf",robot.getX(),
```

```
robot.getY(),robot.getTh());
}

// for backward instruction
if (!strcmp(str,"Backward"))
{
robot.lock();
robot.move(-bStep);
robot.unlock();
while(true)
{
robot.lock();
if(robot.isMoveDone())
{
robot.unlock();
break;
}
robot.unlock();
ArUtil::sleep(50);
}
}

// for left instruction
if (!strcmp(str,"Left"))
{
robot.lock();
head += lHeading;
head %= 360;
robot.setHeading(head);
robot.unlock();

while(true)
{
robot.lock();
if(robot.isHeadingDone())
{
robot.unlock();
break;
}
robot.unlock();
ArUtil::sleep(50);
}
}

// for right instruction
if (!strcmp(str,"Right"))
{
```

```
robot.lock();
head -= rHeading;
head %= 360;
robot.setHeading(head);
robot.unlock();

while(true)
{
robot.lock();
if(robot.isHeadingDone())
{
robot.unlock();
break;
}
robot.unlock();
ArUtil::sleep(50);
}
}

// for Disconnect instruction
if(!strcmp(str,"Disconnect"))
{
printf("Disconnecting from the client.");
break;
}
}
}

// closing and exiting
printf("\n Closing the server");
client.close();
server.close();
Aria::shutdown();
return 0;
}
```

Compilation and execution: The following text command is used for compilation and execution of Tele-operation program

```
g++ -o teleop -I$ARIA/include -L$ARIA/bin -ldl -pthread
-lAria teleop.cpp
```

Listing 11.2. Client source code for tele-operation program.

```java
// filename : Teleop.java
import java.awt.*;
import javax.swing.*;
import java.io.*;
import java.net.*;
import java.awt.event.*;
import java.text.*;
import java.lang.*;
import java.util.*;

public class Teleop extends JFrame implements Ac-
tionListener
{
Container container=null;
Socket socket;
InputStream input;
OutputStream output;
String str[] = {"Move Forward","Move Back","Move
Left","Move Right"};
char mnemonics[] = {'F','B','L','R'};
int bWidth = 120; // button width
int bHeight = 25; // button height
int loc [][] =
{{70,220},{210,220},{140,190},{140,250}};
JLabel x,y,th;
int th1=0;
double x1=0.0,y1=0.0;
JLbl lbl;

public Teleop(String title)
{
super(title);
container = this.getContentPane();
container.setLayout(null);

// for arrow keys
addKeyListener(new KeyAdapter()
{
public void keyPressed(KeyEvent e)
{
double ang=Math.toRadians((double)th1);
try{
if (e.getKeyCode()==e.VK_UP)
{
writeString(output,"Forward");
x1 += 0.1*Math.cos(ang);
y1 += 0.1*Math.sin(ang);
```

```
}
if (e.getKeyCode()==e.VK_DOWN)
{
x1 -= 0.1*Math.cos(ang);
y1 -= 0.1*Math.sin(ang);
writeString(output,"Backward");
}
if (e.getKeyCode()==e.VK_LEFT)
{
writeString(output,"Left");
th1 += 10;
if (th1 == 360)
th1 = 0;
}
if (e.getKeyCode()==e.VK_RIGHT)
{
th1 -= 10;
if (th1 < 0)
th1 += 360;
writeString(output,"Right");
}
x1 = (int)(x1*100);
y1 = (int)(y1*100);
x1 = x1/100;
y1 = y1/100;
x.setText("X :: "+x1);
y.setText("Y :: "+y1);
th.setText("Th :: "+(th1));
lbl.setValues(x1,y1,th1);
lbl.repaint();
}
catch( IOException err) {}
}});

container.requestFocus();

// add all the buttons
for ( int i = 0 ; i < str.length ; i++ )
{
JButton b = new JButton(str[i]);
b.setBounds(loc[i][0],loc[i][1],bWidth,bHeight);
b.setMnemonic(mnemonics[i]);
container.add(b);
b.addActionListener(this);
}
```

```
// add the label for connection
JLabel cLabel = new JLabel("Not Connected");
cLabel.setBounds(200,280,150,30);
container.add(cLabel);

// for x,y,th
x = new JLabel("X :: 0.0");
y = new JLabel("Y :: 0.0");
th = new JLabel("Th :: 0");
x.setBounds(10,250,100,15);
y.setBounds(10,265,100,15);
th.setBounds(10,280,50,15);
container.add(x);
container.add(y);
container.add(th);

// the simulator
lbl = new JLbl();
lbl.setBounds(0,0,399,170);
container.add(lbl);

// for opening the client socket
try
{
socket = new Socket("192.168.0.9",5000);
cLabel.setText("Connected to 192.168.0.9");
}
catch(UnknownHostException e)
{
System.out.println(e);
System.exit(ERROR);
}
catch(IOException e)
{
System.out.println(e);
System.exit(ERROR);
}

//opening the ip/op streams
try
{
output = socket.getOutputStream();
input = socket.getInputStream();
}
catch(IOException e)
{
System.out.println(e);
}
```

```
// adding the window features
addWindowListener(new WindowEventHandler());
setDefaultCloseOpera-
tion(WindowConstants.DISPOSE_ON_CLOSE);
setSize(400,350);
show();
}

// for action performed
public void actionPerformed(ActionEvent ae)
{
String s = ae.getActionCommand();
double tx,ty,ang=Math.toRadians((double)th1);
try{

if (s.equals(str[0]))
{
writeString(output,"Forward");
x1 += 0.1*Math.cos(ang);
y1 += 0.1*Math.sin(ang);
}
if (s.equals(str[1]))
{
x1 -= 0.1*Math.cos(ang);
y1 -= 0.1*Math.sin(ang);
writeString(output,"Backward");
}
if (s.equals(str[2]))
{
writeString(output,"Left");
th1 += 10;
if (th1 == 360)
th1 = 0;
}
if (s.equals(str[3]))
{
th1 -= 10;
if (th1 < 0)
      th1 += 360;
writeString(output,"Right");
}
x1 = (int)(x1*100);
y1 = (int)(y1*100);
x1 = x1/100;
y1 = y1/100;
x.setText("X :: "+x1);
```

```java
y.setText("Y :: "+y1);
th.setText("Th :: "+(th1));
lbl.setValues(x1,y1,th1);
lbl.repaint();
}catch( IOException e) {}
}

// the destructor function
void finalise()
{
try
{
socket.close();
}
catch(IOException e)
{
System.out.println(e);
}
}

// for getting the data from the socket
private String getString(InputStream in) throws IOEx-
ception
{
int c;
int pos =0;
byte buf[]= new byte[1024];

pos = in.read(buf);
if(pos<=0) return null;
String str = new String(buf,0,pos);
return str;
}

// for writing data into socket.
public void writeString(OutputStream o,String s) throws
IOException
{
o.write(s.getBytes());
}

// for window features
class WindowEventHandler extends WindowAdapter
{
public void windowClosing(WindowEvent e)
{
try
```

```
{
writeString(output,"Disconnect");
}
catch(IOException exc){}
System.exit(0);
}
}

// the main function
public static void main(String args[])
{
Teleop frame = new Teleop("Tele-operation");
}

// the simulator
class JLbl extends JLabel
{
public int width = 380,height= 200;
public int x=197,y=97,d=14,th=0;
public int cx=x,cy=y;

public void TLbl()
{
}
public void paint(Graphics g)
{
// the simulator region
g.setColor(Color.white);
g.fillRect(10,10,370,170);
g.setColor(Color.black);
g.drawRect(10,10,370,159);

//the robot at cx,cy
g.drawOval(cx-d/2,cy-d/2,d,d);
int x1 = (int)(d*Math.cos(Math.toRadians(th))/2);
int y1 = (int)(d*Math.sin(Math.toRadians(th))/2);
g.drawLine(cx,cy,cx+x1,cy-y1);
}

//get the Icon width
public int getIconWidth(){return width;}

//get the Icon height
public int getIconHeight(){return height;}
```

```
// set the values.
public void setValues(double x1,double y1,int th1)
{
x1 %= 8;
y1 %= 3;
cx = x + (int)(x1*20);
cy = y - (int)(y1*20);
th = th1;
}
}
}
```

Program log session

```
Syncing 0
Attempting to close previous connection.
Syncing 0
Syncing 1
Syncing 2
Connected to robot.
Name: arcane
Type: Pioneer
Subtype: p2de
Loaded robot parameters from p2de.p

Opened the Server Socket.
Connected to Client.
Client said :: Left 4 4
Client said :: Forward 7 7
64.923000 9.690000 9.228631
Client said :: Forward 7 7
192.831000 32.946000 10.722791
Client said :: Forward 7 7
301.359000 51.357000 8.613389
Client said :: Forward 7 7
407.949000 68.799000 10.195440
Client said :: Right 5 5
Client said :: Backward 8 8
Client said :: Right 5 5
Client said :: Disconnect 10 10
Disconnecting from the client.
Closing the server
Disconnecting from robot.
```

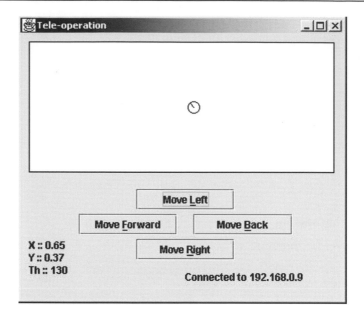

Fig. 11.3. Output of tele-operation program

11.3 Summary

It can be seen from Fig. 11.3 that the simulator consists of four buttons. One can use these four buttons or the mnemonics (Alt + understrike letter). With each action performed the robot in the simulator also moves accordingly. When it crosses the boundary it is again brought back into the region of the simulator. The current position is given by (X, Y) and the direction by theta (Th). Using this program one can run the robot safely from a distance. Here "Direct Motion" commands are used for the tele-operation, but instead "Actions" may be used.

12 A Complete Program for Autonomous Navigation

12.1 Introduction

This chapter discusses the design of a complete navigator program using a client–server architecture for the mobile robot Pioneer 2-DX in a multi-platform system, where the server works on Linux and clients run on the Windows environment. Robot control is achieved through the client–server architecture as shown in Fig. 12.1. The image server program is written in C++ [Swan, 2000; Klander, 2000], and runs on the server for sending the images taken by the framegrabber to the client; and the robot motion server program executes the motion commands on the server and sends low-level commands to the motors. Secondly, the navigator client program is elaborated, which is developed using Java. The program directory layout in the robot's onboard computer is as follows.

```
/home/
    motion/
        Makefile                 the description file for make
        Server                   the motion server executable
        Socket.h                 header file for Socket class
        Socket.cpp               Socket class definition file
        Socket.o                 object file generated after compilation
        ServerSocket.h           header file for ServerSocket
        ServerSocket.cpp         ServerSocket class definition file
        ServerSocket.o           object file generated after compilation
        SocketException.h        header file for socket exceptions
        simple_server_main.cpp   source code for the motion server
        simple_server_main.o     object file generated after compilation
    vision/
        trialserver              Black & white and Color image server
        trialserver.cpp          image server source file
```

trialserver.o	object file generated after compilation
image	RLE encoded image server for **Navigator**
imageserver.cpp	RLE encoded image server code
imageserver.o	object file generated after compilation

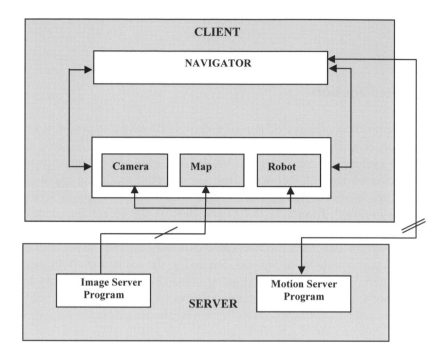

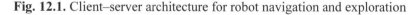
- 8 bit pixel value stream from server to client
- Character based command word stream from client to server and robot state variables from server to client

Fig. 12.1. Client–server architecture for robot navigation and exploration

12.2 The ImageServer Program

The image server program opens a listening socket on port 4325 of the robot's onboard computer. On receiving a request from the navigator client, the program opens the vision system for continuous video and by using a simple run-length encoding algorithm [Jahne, 1997] compresses the black and white image from the left camera and transfers the image data over the network to the client computer. In the navigator client the image is displayed in the upper left corner of the window. The incoming images from the server are continuously displayed. The program encodes the black and white image taken by the framegrabber of the server into the run-length encoded image and transmits over the network using the socket. This program resides in the /home/vision folder of the robot's computer. The source code is available in Listing 12.1 at the website of the book. Telnet session after execution is given below:

Listing of a Telnet session

```
Red Hat Linux Release 7.1 (Seawolf)
Kernel 2.4.2-2.VSBC6 on an i586
login: guest
Last login: Sun Apr 21 15:45:54 from 192.168.0.3
 ~]$ su
 /home/ActivMedia]# cd ../vision
 /home/vision]# image
IEEE 1394 interface open request
1 card(s) found, 2 node(s)
Checking card 0, node 0
Vitana api addr: 78080600
Vendor length is 10
Vendor is: VITANA
Model length is 10
Model is: PixeLINK(tm)
Camera found at node 0 0: VITANA PixeLINK(tm)
Camera ISO bandwidth needed: A10
Max_Image_Size_Inq: 05080408
Unit_Size_Inq: 00080008 (0)
Image_Size_Inq: 05000400 (0)
Frame_Rate_min: 0000000E (0)
Frame_Rate_max: 000000A0 (0)
Frame_Rate_def: 00000011 (0)
Flags: 00000000 (0)
PCS2112 ver 0x30
Imager reset starting...
Imager reset succeeded
Imager ready.
Camera ISO speed set to 400 Mb/sec
Camera ISO parameters: 2000000
Opened frame grabber.
Size: 320/1280 240/960
Image server listening on 0.0.0.0:4325
```

12.3 The MotionServer Program

This program controls the robot's movements, obstacle detection and gripper functions. The socket communication in this program is encapsulated in `Socket` and `SocketServer` C++ classes. These classes implement methods for handling low-level LINUX socket communication [Mitchell, 2001]. The `Socket` class implements all the basic socket operations (like bind, listen, accept, connect, write and recv) and the `SocketServer` class adds the possibility of exceptions that may occur during the lifecycle of a socket. In this section, the program opens a connection to the robot or the robot simulator and then waits on port 4040 for a connection from the Navigator client. On negotiating a successful connection with the navigator client, the program starts an infinite loop in which it sends the client information about the robot's state, such as the robot's position, heading, translational and rotational velocity and battery voltage. The socket handling the communication is set as non-blocking so that a read function on the socket does not block the program. This is done because the data or command sent from the client to the motion server is of an asynchronous nature. And if the program is blocked during the read operation, the client will not be provided with the robot's state information. The client program sends specific commands in the form of special strings. The program then interprets the commands and if a match is found the associated function is executed. The list of valid commands is given in Table 12.1. The listing of various program codes for the robot motion server is available in Listing 12.2 at the website of the book. The Telnet session is given below.

Table 12.1. Lists of commands for robot motion

HEAD	The client sends a heading value followed by this command. The motion server on receiving the command sets the robot's heading accordingly.
MOVE	The client specifies the distance to be moved followed by this command. The motion server on receiving the command issues a move command to the robot for the given distance.
HALT	This brings the robot to a standstill, discarding all motion commands already executing.
UMOV	Moves the gripper lift Up.
DMOV	Moves the gripper lift Down.
GOPN	Opens the gripper paddle.
GCLS	Closes the gripper paddle.
STOP	Stops the gripper.

SOFF	Turns off BOTSPEAK's commentary.
SONN	Turns on BOTSPEAK's commentary.
FRWD	Moves the robot forward by 15 cm.
MBCK	Moves the robot back by 10 cm.
TLFT	Turns the robot left by 10 degrees.
TRGT	Turns the robot right by 10 degrees.
WNDR	Starts wander action on the robot.
SWND	Stops wander action if already running.
PFDR	Starts path-finder action on robot.
SPFR	Stops path-finder action on robot, if running.
OBTR	Starts object-tracking action on robot.
SOTR	Stops object tracking action on robot, if running.

A Telnet session for running the motion server

```
Red Hat Linux release 7.1 (Seawolf)
Kernel 2.4.2-2.VSBC6 on an i586
login: guest
Last login: Sun Apr 21 15:47:46 from 192.168.0.3
aumix:  error opening mixer
aumix:  error opening mixer
 ~]$ su
 /home/ActivMedia]# cd ../motion
 /home/motion]# nohup startx &
 1] 841
 /home/motion]# nohup: appending output to `nohup.out'

 /home/motion]# export DISPLAY=localhost:0
 /home/motion]# xhost +localhost
localhost being added to access control list
 /home/motion]# server
running....
Botspeak server: ret=0 from eciSetParam
Botspeak server: ConnectToEngine invoked
             IBM ViaVoice Speech Recognizer

(C) Copyright International Business Machines Corp.
1991-1999.
All Rights Reserved
Licensed Materials - Property of IBM
```

```
U.S. Government Users. RESTRICTED RIGHTS -
Use, Duplication, or Disclosure restricted by GSA ADP
Schedule Contract with IBM Corporation.

Recognizer: initializing wsi ... pap ... dec ...
ready.

Botspeak server: ConnectToEngine: SmConnect() rc = 0
Botspeak server: Initializing ILU
Botspeak server: Server instance published.
Botspeak server: Its SBH is
"ilusbh:BotSpeak/iface_Obj;ilut-0X1.FFA14080F4B60P+0
jCJjhsJx98pxHkmmo7-
5CEPQUOh;sunrpc@sunrpcrm=tcp_192.168.0.9_1027".
Botspeak server: Connected to speech engine.
BotSpeak: No callback defined for SmNfocusGrantedCall-
back
MicOffCB: rc = 5
Syncing 0
Syncing 1
Syncing 2
```

Connected to robot.

```
Name: Burla_1473
Type: Pioneer
Subtype: p2de
Loaded robot parameters from p2de.p
Gripper:  queried, using General IO.
```

The BOTSPEAK server requires that X-Windows be running on the client. The nohup command is used to start X-Windows (startx) so that even after the completion of the Telnet session the child, i.e. the X-server, is not terminated. After starting the X-server, the display variable is exported such that using xhost the X-server on the robot's computer can host the display that the BOTSPEAK server intends to open in the Telnet session.

12.4 The Navigator Client Program

This program is written in Java, because it supports a networking capability and provides built-in classes for window-based display [Java, 2002]. The program has been broken up into several class files each for handling a different task. The program is multi threaded, and spawns a thread for receiving the robot's state information and reflecting it in the client display. The entire program has been broken up into the following classes.

- **Client:** This handles socket communication with the motion server and the image server.

- **Camera:** This handles the reception and display of the run-length encoded image received from the image server.

- **Map:** This displays the robot's position and range sensor information in 2D Cartesian coordinate display.

- **Robot:** This encapsulates the robot properties for display in the above coordinate system.

- **Navigator:** This integrates all the classes and implements the user interface with the program. It also receives the robot's state information on a separate thread.

The navigator is the main class and brings together all the functionalities of the classes. This program spawns two threads one to receive the image from the image server and the other for receiving the state information of the robot. On the main thread the program listens to user events and accordingly commands the motion server. The various program codes of the navigator is available in Listing 12.3 of the website of the book. The navigator programs can be executed from the command prompt by issuing the following commands:

```
C:\> cd Navigator
C:\Navigator\> set PATH=%PATH%;C:\jdk1.4\bin
C:\Navigator\> java Navigator
```

Fig. 12.2. The robot navigator program window

The navigator program must be launched after starting both the image server and the motion server programs. These server programs can be launched from separate Telnet sessions with the robot's onboard computer. The procedural details have been discussed with the individual programs. On launching the navigator client program, connection is established with the image server and the motion server. The front end of the Navigator Client program is shown in Fig. 12.2. The upper left sub-window gives a continuous display of the run-length encoded black and white image received from the left camera of the stereo rig. The right half of the screen displays the bird's-eye view of the robot along with the sonar sensor readings as a red dot around the robot. Robot control functions are available in the left lower panel. The Open, Close, Up, Down, Ready and Stop button in the first row control the gripper of the robot. In navigator, the robot can be operated in three different modes and these are available in the Navigate drop down menu.

Fig. 12.3. Bird's-eye view of the robot's surrounding

- **Tele-operate:** This is the default mode where the keyboard keys are used to control the robot. To activate this mode one must pay attention to the right display sub-window. This is achieved by a mouse click in the subwindow. The following keys are used to operate the robot:

Up arrow (↑)	move the robot forward
Down arrow (↓)	move the robot back
Left arrow (←)	turn the robot left
Right arrow (→)	turn the robot right
"o"	open gripper paddles
"c"	close gripper paddles
"u"	lift up the gripper paddles
"d"	move down the gripper paddles

- **Wander:** This activates the wander activity in the robot. The Go button is used to launch this mode and the Halt button stops it. In this mode the robot autonomously wanders around avoiding obstacles on its own and sends the sonar readings and video image from the left camera. This is used to take the sensory readings of the robot.

- **Mapper:** In this mode a path for the robot is specified by clicking at points, that the robot is supposed traverse. A line on the display indicates the projected path of the robot. The path can be closed at the origin of the robot by a right click. The Go button starts the procedure and the robot starts tracing the path gives by the yellow lines. On reaching the destination the robot stops. The Halt button may be used to terminate the procedure before completion. If the robot finds an unavoidable obstacle it terminates the procedure.

The text box named Speak activates the robot's speech-synthesis system. The text in the text box will be read aloud by the robot's BOTSPEAK system using the IBM VIAVOICE speech engine. The Commentary checkbox enables or disables online commentary of the actions taken by the robot. If the "ready" button is activated, the robot can grab and lift an object, whenever it finds an object between its open paddles. On the control panel, lower left side, one can find state variable indicators such as position (x, y), heading, velocity (translational and rotational) and battery voltage. A bird's-eye view of the robot's surroundings is shown in Fig. 12.3 and the image sequence is shown in Fig. 12.4.

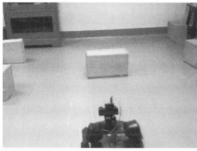

Image Sequence 1

Image Sequence 2

Image Sequence 3

Image Sequence 4

Image Sequence 5

Image Sequence 6

Image Sequence 7

Image Sequence 8

Fig. 12.4. Sequence of images of robot navigation

12.5 Summary

The chapter describes the development of a complete client–server archi-tecture for the mobile robot in a multiplatform network for the navigation and exploration of Pioneer 2-DX. It is useful to add various modules in the same program and test their algorithms, with minimum changes, which releases the burden of designing the front end of the client–server architecture.

13 Imaging Geometry

13.1 Introduction

Visual perception is undoubtedly one of the most precious and trustworthy sense organs of human beings for understanding the environment. In the recent past, researchers have added this invaluable faculty to machines in addition to their intelligence. Machine vision refers to the viewing or sensing of the environment by the computer, allowing it to synthesize information from the imagery of the concerned scene, analyze it, and finally carry out various interpretations or make decisions. Computer vision employs 3D imaging techniques, which differ remarkably from classical imaging in such a way that it recovers the depth information, i.e. the third dimension, by various techniques. In fact, it is a formidable task to emulate the human visual system in machines, since it requires a detail understanding of the imaging process. It finds extensive application in navigation and path planning of mobile robots, tracking and targeting in air missiles and defense systems, the manufacturing environment, disposal of toxic waste in nuclear power plants, etc. For accomplishing this task the robot has to be equipped with cameras to obtain visual information about its neighborhood.

13.2 Necessity for 3D Reconstruction

While programming the mobile robot to carry out navigational tasks autonomously, it is always necessary to explore the surroundings. Recently the problem of exploring an unknown environment has received considerable attention from the computer science and AI community. Before exploration, it is assumed that the environment is populated with a polygonal obstacle and the robot has to determine its position with respect to a global frame of reference. In practice, it is quite difficult to estimate precisely the position of a mobile robot with respect to an arbitrary frame of reference. Therefore, in robotic systems an odometry system is used to determine the robot's global position but this suffers from the problem of cumulative errors as the robot moves further from its starting position. In order to avoid the cumulative errors, many localized systems require installing a set of beacons at known locations in the robot's workspace. But this is not viable for robots working in indoor environments. To do some complicated task, such as a pick and placement job in a flexible manufacturing environment, searching for leaking barrels of toxic waste in nuclear plants, surface moving where the surface is not planar and familiar, vision-based 3D exploration is necessary. However 3D scene recovery from 2D planar images remains a challenging task; even today, it is overwhelmed with problems.

13.3 Building Perception

Perception, as discussed in Chap. 1, constructs higher-level knowledge from relatively lower level data or knowledge. Generally the noise-free information is stored in LTM by the state of acquisition. The state of perception employs reasoning tools on the information recorded in LTM and thus derives new rules for subsequent planning and coordination problems. A mobile robot constructs its surrounding map by sensing information around it and preprocessing that information at the state of acquisition. In a 2D planning problem, the boundaries of the obstacles are generally sensed by ultrasonic sensors or laser range-finders. It is always assumed that the top of the obstacles is at a level higher than the mounting point of the sensors. A 3D planning problem, on the other hand, requires keeping track of the obstacle surfaces and their heights as well. To extend the 3D information, generally additional cameras are employed. These cameras are mounted on a pan-tilt platform, which is fixed with the mobile robot. When more than one camera are used for determining the third dimension of the obstacles, it is called stereo vision.

For constructing a 2D world map, the robot has to move around each obstacle. Starting from a given location the robot moves around each obstacle, until all obstacles are visited. A two-dimensional world map for the robot is then built up with the visited obstacles. If the sensory information recorded in the Long Term Memory (LTM) is not completely free from noise, then noise has to be eliminated first.

13.3.1 Problems of Understanding 3D Objects from 2D Imagery

One of the limitations of machine vision is that the imaging process is interpreted from the information on a 2D image relating to the 3D world. But this can be modeled with the knowledge of the physical process of image formation, i.e. how the 3D world scene is projected onto the image plane of the camera giving rise to the 2D imagery. This basically constitutes the basis of *perspective projection*, which will be covered later in this chapter. It is not possible to recover the 3D spatial geometry from a single image. However if we have multiple images of the scene at hand, taken from different viewing angles and positions, then it is possible to extract the 3D geometric information by a combination of these images, which is known as stereo-vision.

The camera projects the 3D world onto its image plane. The image formed in the image plane of the camera is essentially 2D in nature. As mentioned earlier, it is necessary to have at least two 2D images to interpret the 3D information contained in them. But another difficulty comes in between in merging the two images as there may be many uncertainties involved in this imaging. The first one is the uncertainty regarding the position and orientation of the various geometrical primitives in the image planes, which give rise to uncertainty in estimation of the 3D features. Secondly, uncertainty in camera parameters such as focal length and random measurement errors also adds uncertainty in the measurement process. Little relevant work has been reported in the area of 3D exploration of the environment or 3D object reconstruction. Reconstruction of 3D features by structured light and least squares estimation techniques have been explained in by Haralick and Shapiro [Haralick, 1993]. Asada [Asada, 1990] has fused sensory/camera images and a sonar range scanner for reconstruction of a world map.

13.3.2 Process of 3D Reconstruction

The process of 3D reconstruction starting from the raw image received through the camera is shown by the block diagram Fig. 13.1. The basic tasks can be divided into two steps:

1. To extract meaningful affine geometric information about the environment from the raw image received through the camera(s) and model the uncertainty in the process of measurement.

Image Data received through Camera

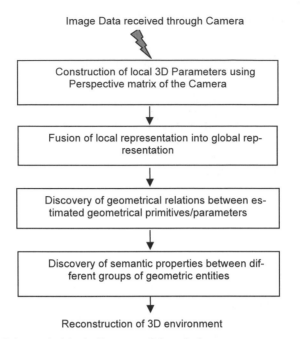

Fig. 13.1. Schematic block diagram of the whole process

2. To fuse the local representation into a global representation and model
   the uncertainty of fusion as well as the displacement of the robot.

After the geometric primitives are obtained, the local primitives are to be
fused to obtain a global 3D reconstruction of the scene/object, which is
represented in the block diagram in Fig. 13.2. Here the objective is to
eliminate uncertainties resulting from the measuring process, first in the
process of extraction of 2D image features and secondly in the 3D reconstruction process, using a Kalman filter.

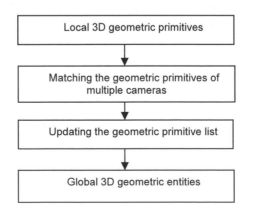

Fig. 13.2. Block diagram for estimation of global 3D information

13.4 Imaging Geometry

The process of 3D imaging involves capturing the image in the 3D world. The primary task is to develop an imaging system that could emulate the human visual process. Let us discuss how to recover 3D information from multiple 2D images obtained from cameras at different viewing positions. 3D imaging can be classified into (i) planar imaging, where a plane is imaged; (ii) surface imaging, which involves surfaces, and (iii) imaging 3D objects, which deals with 3D objects. Planar imaging is the fundamental one, which is available in almost all image processing books, and surface imaging, which deals with the analysis of the surface of any non-planar object, will be covered in a separate chapter in this book.

13.4.1 Image Formation

Energy from the 3D world is converted into two-dimensional entities called images by imaging systems such as a camera or electro-optical sensors. Once the image is acquired, the 2D images are spatially sampled and quantized after which the digital image has been digitized both in spatial coordinates and brightness. Image formation involves *radiometric* as well as *geometric* aspects. Radiometric aspects link the radiance of the object to the irradiance of the image plane, which depends on the amount of radiation emitted by the object and that collected by the imaging system and finally received by the sensor. Each point in the image has a characteristic gray level (intensity value) corresponding to the point in the 3D world. It is represented by luminance and reflectance components. These components vary over the entire range of the image and can be made use of in carrying out low level imaging such as detection of edges, lines, shade, texture, etc.

The second aspect of image formation involves geometric aspects, which relate the position of the object in 3D space to its position in the image plane. The basic difficulty of image formation results from the fact that the 3D world is projected onto the 2D image plane. Interpreting the information of the 3D world requires understanding the way the projection takes place. Through a careful modeling of the imaging process, we can effectively model the camera as a pinhole and the mapping/projection of the 3D world onto the image plane is then referred to as perspective projection.

Here only a single ray from a given point of the object in the 3D world is allowed to pass through the pinhole and form a point image on the image plane. This effectively and adequately models the imaging process and is commonly used in imaging systems. Let us first discuss the principle of perspective projection in one dimension.

13.4.2 Perspective Projection in One Dimension

To illustrate the concept of perspective projection, consider a camera taking one-dimensional pictures in a two-dimensional world. Here for the sake of simplicity, the coordinate systems related to the camera and the world coordinate system coincide.

As shown in Fig. 13.3, the camera lens is at the origin and points directly to the Y-axis. In order to keep the image in a positive orientation, let us assume that the image line is at a distance f in front of the camera lens and that the lens projects towards it. This avoids the confusion of left–right reversal in an image behind the lens. The image line is parallel to the X-axis.

In accordance with geometric ray optics, the ray will focus point (r, s) on to the image line, which is a line parallel to the X-axis and at a distance f directly in front of the lens. The position of the line is determined by where the line from (r, s) to the origin intersects the image line. Hence perspective projection has coordinates $(rf/s, f)$ in the original two-dimensional coordinate system. Here it can be seen that the relationship between the point and its image bears a nonlinear relationship. However, the numerator and denominator of rf/s are linear combination of r and s. This indicates that the camera transforms the point (r, s) to the image point I by linear transformation T in projective coordinates. This can be illustrated by using a homogeneous coordinate system which will be cov-

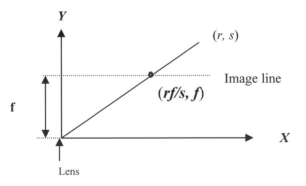

Fig. 13.3. Two dimensional perspective projection

ered in the next section. The point (r, s) is represented as $(r, s, 1)$ in the homogenous coordinate system. The first linear transformation translates the point $(r, s, 1)$ down the Y-axis by a distance f. The second transformation takes the perspective transformation to the image line. Hence,

$$\begin{pmatrix} ku \\ kv \end{pmatrix} = \begin{pmatrix} 1 & 0 & 0 \\ 0 & \frac{1}{f} & 1 \end{pmatrix} \begin{pmatrix} 1 & 0 & 0 \\ 0 & 1 & -f \\ 0 & 0 & 1 \end{pmatrix} \begin{pmatrix} r \\ s \\ 1 \end{pmatrix} \tag{13.1}$$

With this understanding of perspective projection, let us discuss the process of 3D perspective projection.

13.4.3 Perspective Projection in 3D

The camera is modeled as a pinhole with an optical center and an image plane which is usually perpendicular to the optical axis, the block diagram of which is represented in Fig. 13.4. Imaging systems, when used to image a scene from different directions and positions, make use of several coordinate systems, i.e. a world or user selected coordinate system which is related to the observed scene, and the camera coordinate system which is usually centered at the optical center and whose Z-axis is aligned along the optical axis of the camera. The image plane centric coordinate system, which is aligned with the camera coordinate system, shifts a distance f from the optical center. The sensor-based coordinate system is attached to the sensor and depends on the arrangement of the pixel matrix.

There are several frameworks for defining the configuration of the various coordinate systems. Here let us assume the optical center to be the origin of the camera coordinate system and the projection plane to be situated at $(0, 0, -f)$, i.e. at a distance of f behind the optical center. Here f is known as the principal distance or camera focal length. For the sake of

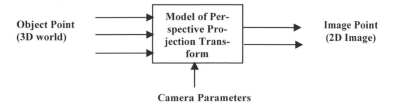

Fig. 13.4. Imaging geometry in 3D space

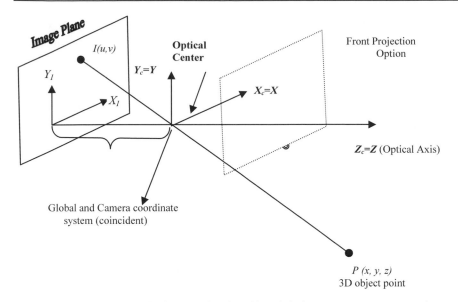

Fig. 13.5. Camera model of 3D projection (the global cooroinate system and camera coordinate systems are assumed to be the same here)

simplicity, at first the world coordinate system is taken to align with the camera coordinate system.

In Fig. 13.5, a point having spatial coordinates (x, y, z) is mapped onto its image I(u, v) on the image plane. Using the principle of perspective projection

$$\frac{x}{z} = -\frac{u}{f} . \Rightarrow u = -\frac{f}{z}x$$

$$\frac{y}{z} = -\frac{v}{f} . \Rightarrow v = -\frac{f}{z}y \qquad (13.2)$$

Properties of the 3D Projection System

1. It can be seen that the two world coordinates parallel to the image plane are scaled by a factor f/z. Thus image coordinates contain ratios of world coordinates.
2. The imaging model for perspective projection also maps straight lines in the world onto straight lines on the image plane. However, perspective

projection does not preserve distances between points, nor their ratios. Thus if A, B, C are collinear in 3D space, then

$$AC/BC \neq ac/bc \tag{13.3}$$

where AC and BC represent the distance between points A and B, B and C, etc. while ac, bc, denote the distances between the corresponding projected points on the image plane. It is also seen from Fig. 13.5 that an object point located anywhere on the line segment PO will be projected onto the same point I. This shows that perspective projection is a many-to-one transformation.

3. Equation (13.2) is essentially nonlinear in nature since it involves division by z. However, the numerator and denominator are linear combinations. It is usually desired to express the mapping from 3D to 2D by means of a linear transformation. To effect this linear transformation, requires the use of a homogenous coordinate system. The homogenous coordinates for the physical point (x, y, z) in 3D space are represented by the 4×1 vector (kx, ky, kz, k) where k is an arbitrary scalar. To convert the homogenous system back to the physical coordinate system, we divide all components by k and delete the row containing it.

At this point, we can put the 3D–2D transformation in a linear matrix form given by (13.4).

$$\begin{pmatrix} ku \\ kv \\ kw \\ k \end{pmatrix} = \begin{bmatrix} 1 & 0 & 0 & 0 \\ 0 & 1 & 0 & 0 \\ 0 & 0 & 1 & 0 \\ 0 & 0 & -1/f & 0 \end{bmatrix} \begin{bmatrix} x \\ y \\ z \\ 1 \end{bmatrix} \tag{13.4}$$

This is the representation of the 2D point in the image plane located at $(0, 0, -f)$. Hence the value of w will be $kw/w = z/(-z/f) = -f$ as verified. Now let us try to recover the 3D object point from its coordinates in the image plane by taking the product of the inverse of the linear transformation matrix and the image plane coordinates. Then we will definitely arrive at erroneous results, or ambiguity. This is because of the many-to-one transformation characteristic of the perspective matrix. Knowledge of at least one 3D coordinate is essential for reconstruction of the 3D point. If we try to represent the image plane coordinates w.r.t. a coordinate system centered on the image plane as shown in Fig. 13.6, then it is required to translate the coordinate system from the center of perspectivity to the image plane by a distance f along the optical axis. The point will now be represented by two coordinates (u, v) and thus the homogenous representation will be (ku, kv, k) with $w = 0$. It is thus necessary to delete the third row in the linear transformation matrix, which is also known as the perspective matrix and it takes the form shown in (13.5) below. It is now seen that the perspective matrix becomes non-invertible.

$$\begin{pmatrix} ku \\ kv \\ k \end{pmatrix} = \begin{pmatrix} 1 & 0 & 0 & 0 \\ 0 & 1 & 0 & 0 \\ 0 & 0 & -1/f & 0 \end{pmatrix} \begin{pmatrix} x \\ y \\ z \\ 1 \end{pmatrix} \tag{13.5}$$

The above linear transformation matrix may be visualized in terms of two transformations i.e. one for translating the coordinate system from the center of perspectivity to the image plane-centric system, and the other for taking the perspective projection. Hence the transformation can be represented by (13.6).

$$\begin{pmatrix} ku \\ kv \\ k \end{pmatrix} = \begin{pmatrix} 1 & 0 & 0 & 0 \\ 0 & 1 & 0 & 0 \\ 0 & 0 & -1/f & 1 \end{pmatrix} \begin{pmatrix} 1 & 0 & 0 & 0 \\ 0 & 1 & 0 & 0 \\ 0 & 0 & 1 & f \\ 0 & 0 & 0 & 1 \end{pmatrix} \begin{pmatrix} x \\ y \\ z \\ 1 \end{pmatrix} \tag{13.6}$$

Analogously, if we take the front projection the perspective relation takes the form of an equation as shown in equation (13.7).

$$\begin{pmatrix} ku \\ kv \\ k \end{pmatrix} = \begin{pmatrix} 1 & 0 & 0 & 0 \\ 0 & 1 & 0 & 0 \\ 0 & 0 & 1/f & 1 \end{pmatrix} \begin{pmatrix} 1 & 0 & 0 & 0 \\ 0 & 1 & 0 & 0 \\ 0 & 0 & 1 & -f \\ 0 & 0 & 0 & 1 \end{pmatrix} \begin{pmatrix} x \\ y \\ z \\ 1 \end{pmatrix} \tag{13.7}$$

It has been assumed so far that the camera and the world coordinate system coincide with each other. In this case, however, when it is required to estimate the 3D features w.r.t. a user selected world coordinate system, it is a cumbersome task to take the camera system as the reference for each view position. Hence we need to dissociate the camera and world coordinate system, when we take several images of the 3D scene from different viewing position.

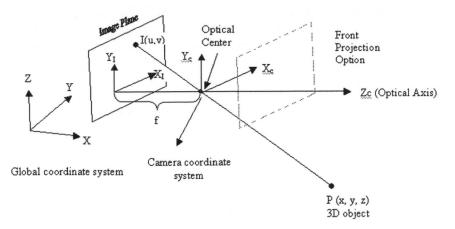

Fig. 13.6. Generalized camera model of 3D projection

In order to consider perspective projection, in such cases it is first required to align the world coordinate system with the camera coordinate system. This can be achieved by performing a translation of the global/world coordinate system to the origin of the camera system by a translation vector denoting the directed distance between them. Then a series of rotations is performed about suitably chosen axes so that the world system coincides with the camera system. The overall process can be summed up by a relation (13.8).

$$\hat{A}_c = \hat{A}' = R(\hat{A} - T) \tag{13.8}$$

where $\hat{A}$ represents the global system (X, Y, Z), $\hat{A}'$ represents the transformed global system, $\hat{A}_c$ represents the camera system $(X_c \ Y_c, Z_c)$ and T and R represent translation and rotation vectors.

The process of aligning two coordinate systems is discussed in the next section.

13.5 Global Representation

The construction of the global map requires a common frame of reference, considering the camera coordinate system, which is a cumbersome process. This is because the camera coordinate system goes on changing from position to position and thus it is necessary to take into account the relative changes in the camera position at each location. Hence a user defined coordinate system is used, which is also called the global system. All measurements are then estimated w.r.t. the global system. This ensures convenience in constructing the world/global map by fusing the multiple-image information from several local maps. The camera coordinate system shown in Fig. 13.7 is attached to the robot and can take multiple images of the scene from different positions and directions.

In order to have a global representation, it is necessary to establish a relationship between the maps sensed by camera, i.e. the camera coordinate system and the global coordinate system. It is necessary to bring about a series of transitions of the coordinate system pertaining to the corresponding frames of reference. The first step is the transition of the camera-based coordinate system to the robot coordinate system, for transforming the sensor map to the virtual map. Let (x_c, y_c, z_c), be the displacement vector of the origin of the vehicle based coordinate system w.r.t. the camera-based coordinate system. Then in the estimation, the origin of the camera-based

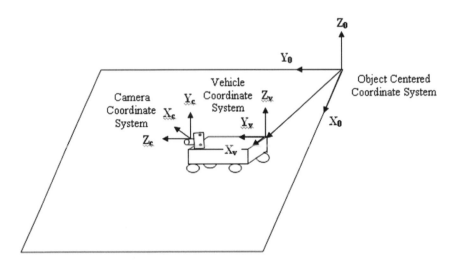

Fig. 13.7. Relationship between different coordinate systems

coordinate system is to be shifted by the above translation vector so that it coincides with the origin of the robot coordinate system, denoted by (13.9)

$$\begin{pmatrix} X'_c \\ Y'_c \\ Z'_c \end{pmatrix} = \begin{pmatrix} X_c - x_c \\ Y_c - y_c \\ Z_c - z_c \end{pmatrix} \tag{13.9}$$

where X_c, Y_c, Z_c corresponds to the axes of the camera-based coordinate system. The alignment of the two system axes after translation is shown in Fig. 13.8.

Next the new coordinate system is aligned with the robot coordinate system, so that a relation is established between the camera-based coordinate system with the robot coordinate system, which is done by a sequence of rotations about properly chosen axes.

Step 1: Rotation about the Z_c' axis by ϕ_1: In this case the $X_c' - Y_c'$ plane is rotated about the Z_c' axis by ϕ_1 where ϕ_1 is the angle between X_c' and the $X_v - Z_c'$ plane measured along the $X_c' - Y_c'$ plane. The new coordinate system is represented by (13.10) and shown in Fig. 13.9.

$$\begin{pmatrix} X''_c \\ Y''_c \\ Z''_c \end{pmatrix} = \begin{pmatrix} \cos\phi_1 & \sin\phi_1 & 0 \\ -\sin\phi_1 & \cos\phi_1 & 0 \\ 0 & 0 & 1 \end{pmatrix} \begin{pmatrix} X'_c \\ Y'_c \\ Z'_c \end{pmatrix} \tag{13.10}$$

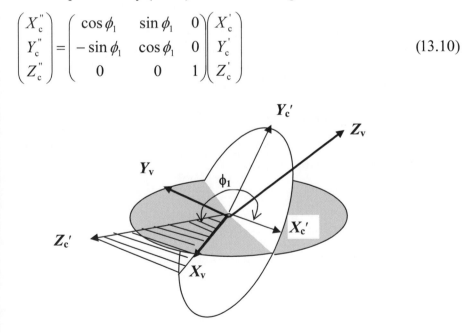

Fig. 13.8. Orientation of camera axes after translation to the robot coordinate system. Y_v and X_v lie on the black plane, while Z_v is perpendicular to it. X_c' and Y_c' lie on the white plane while Z_c' is perpendicular to it. X_c' will be rotated to $X_v - Z_c'$ plane

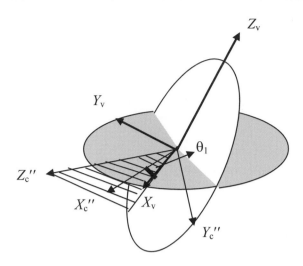

Fig. 13.9. Orientation of camera axes after step 1. Y_v and X_v lies on the black plane, while Z_v is perpendicular to it. Y_c'' lies on the white plane while Z_c'' and X_c'' are perpendicular to it. Now the $X_c'' - Z_c''$ plane will be rotated by an angle θ_1 in order to align X_c'' with X_v

Step 2: Rotation about the Y_c'' axis by θ_1: Here the $X_c'' - Z_c''$ plane is rotated about the Y_c'' axis by an angle θ_1, where θ_1 is the angle between X_c'' and X_v along the $X_c'' - Z_c''$ plane. Thus the new coordinate system is represented by (13.11) and shown in Fig. 13.10.

$$\begin{pmatrix} X_c''' \\ Y_c''' \\ Z_c''' \end{pmatrix} = \begin{pmatrix} \cos\theta_1 & \sin\theta_1 & 0 \\ 0 & 0 & 1 \\ -\sin\theta_1 & \cos\theta_1 & 0 \end{pmatrix} \begin{pmatrix} X_c'' \\ Y_c'' \\ Z_c'' \end{pmatrix} \qquad (13.11)$$

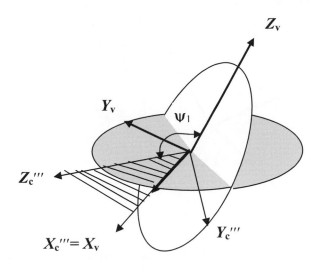

Fig. 13.10. Orientation of camera axes after step 2. Y_v and X_v lie on the black plane, while Z_v is perpendicular to it. Y_c''' lies on the white plane while Z_c''' and X_c''' are perpendicular to it. X_c''' is seen to be aligned with X_v. Now the $Y_c'''- Z_c'''$ plane will be rotated by an angle ψ_1 in order to align Z_c''' with Z_v. Then the Y_c''' will be automatically aligned with Y_v

Step 3: Rotation about the X_c''' axis by ψ_1: Here the $Y_c'''-Z_c'''$ plane is rotated about the X_c''' axis by ψ_1, where ψ_1 is the angle between Z_c''' and Z_v measured along the $Y_c'''-Z_c'''$ plane. Thus the new coordinate system is denoted by (13.12) and shown in Fig. 13.11.

$$\begin{pmatrix} X_c^{iv} \\ Y_c^{iv} \\ Z_c^{iv} \end{pmatrix} = \begin{pmatrix} 0 & 0 & 1 \\ \cos\psi_1 & \sin\psi_1 & 0 \\ -\sin\psi_1 & \cos\psi_1 & 0 \end{pmatrix} \begin{pmatrix} X_c''' \\ Y_c''' \\ Z_c''' \end{pmatrix} \tag{13.12}$$

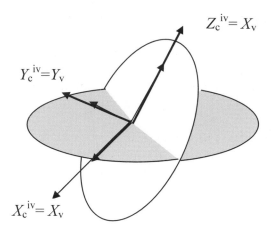

Fig. 13.11. Orientation of camera axes after step 3. The camera and the robot coordinate systems are now aligned after one translation and three subsequent rotations

Now, it can be easily seen that the camera coordinate system can be translated to the robot coordinate system by the relationship (13.13).

$$
\begin{pmatrix} X_v \\ Y_v \\ Z_v \end{pmatrix} = \begin{pmatrix} X_c^{iv} \\ Y_c^{iv} \\ Z_c^{iv} \end{pmatrix}
$$

$$
= \begin{pmatrix} 0 & 0 & 1 \\ \cos\psi_1 & \sin\psi_1 & 0 \\ -\sin\psi_1 & \cos\psi_1 & 0 \end{pmatrix} \begin{pmatrix} \cos\theta_1 & \sin\theta_1 & 0 \\ 0 & 0 & 1 \\ -\sin\theta_1 & \cos\theta_1 & 0 \end{pmatrix}
$$

$$
\begin{pmatrix} \cos\phi_1 & \sin\phi_1 & 0 \\ -\sin\phi_1 & \cos\phi_1 & 0 \\ 0 & 0 & 1 \end{pmatrix} \begin{pmatrix} X_c - x_c \\ Y_c - y_c \\ Z_c - z_c \end{pmatrix} \tag{13.13}
$$

13.6 Transformation to Global Coordinate System

Global representation can be obtained from the robot coordinate system by establishing a relationship between the vehicle-based coordinate system and the object-centered coordinate system. This can be done by the following two steps.

Step 1: Translation: The robot-based coordinate system is first translated by the translation vector (x_v, y_v, z_v) so that the origin of the new system coincides with the global coordinate system, which is denoted by equation (13.14). After which a series of rotation is carried out for the alignment of the two coordinate systems.

$$\begin{pmatrix} X'_v \\ Y'_v \\ Z'_v \end{pmatrix} = \begin{pmatrix} X_v - x_v \\ Y_v - y_v \\ Z_v - z_v \end{pmatrix} \tag{13.14}$$

Step 2: Rotation about the Z_v' axis by ϕ_2: This rotation is accomplished by rotation of the $X_v'-Y_v'$ plane about the Z_v' axis by ϕ_2, where ϕ_2 is the angle between the Y_v' and Y_0 axes. It is assumed here that the Z axis of both coordinate systems coincides (assume the floor to be the X_0-Y_0 plane). Thus the following relationship is obtained

$$\begin{pmatrix} X''_v \\ Y''_v \\ Z''_v \end{pmatrix} = \begin{pmatrix} X_0 \\ Y_0 \\ Z_0 \end{pmatrix} = \begin{pmatrix} \cos \phi_2 & \sin \phi_2 & 0 \\ -\sin \phi_2 & \cos \phi_2 & 0 \\ 0 & 0 & 1 \end{pmatrix} \begin{pmatrix} X'_v \\ Y'_v \\ Z'_v \end{pmatrix}$$

i.e.

$$\begin{pmatrix} X_0 \\ Y_0 \\ Z_0 \end{pmatrix} = \begin{pmatrix} \cos \phi_2 & \sin \phi_2 & 0 \\ -\sin \phi_2 & \cos \phi_2 & 0 \\ 0 & 0 & 1 \end{pmatrix} \begin{pmatrix} X_v - x_v \\ Y_v - y_v \\ Z_v - z_v \end{pmatrix} \tag{13.15}$$

The relationship between the camera and the robot coordinate system was obtained in (13.13) and the relationship between the robot and the global coordinate systems by (13.15). Thus it is possible to obtain a relationship between the camera and object-centered coordinate system by combing these two equations.

Once the alignment is completed, the perspective projection is carried out by summing up all the processes as follows:

$$
\begin{pmatrix} ku \\ kv \\ k \end{pmatrix} = \begin{pmatrix} 1 & 0 & 0 & 0 \\ 0 & 1 & 0 & 0 \\ 0 & 0 & -1/f & 1 \end{pmatrix} \begin{pmatrix} 1 & 0 & 0 & 0 \\ 0 & 1 & 0 & 0 \\ 0 & 0 & -1 & f \\ 0 & 0 & 0 & 1 \end{pmatrix}
$$

$$
\begin{pmatrix} 1 & 0 & 0 & 0 \\ 0 & \cos\theta & -\sin\theta & 0 \\ 0 & \sin A & \cos A & 1 \\ 0 & 0 & 0 & 1 \end{pmatrix} \begin{pmatrix} \cos\phi & & \sin\phi & 0 \\ 0 & 1 & 0 & 0 \\ -\sin\phi & 0 & \cos\phi & 0 \\ 0 & 0 & 0 & 1 \end{pmatrix}
$$

$$
\begin{pmatrix} \cos\psi & -\sin\psi & 0 & 0 \\ \sin\psi & \cos\psi & 0 & 0 \\ 0 & 0 & 1 & 0 \\ 0 & 0 & 0 & 1 \end{pmatrix} \begin{pmatrix} 1 & 0 & 0 & -x_0 \\ 0 & 1 & 0 & -y_0 \\ 0 & 0 & 1 & -z_0 \\ 0 & 0 & 0 & 1 \end{pmatrix} \begin{pmatrix} X \\ Y \\ Z \\ 1 \end{pmatrix}
$$

(13.16)

where (x_0, y_0, z_0) represents the displacement vector of the camera coordinate system with respect to a user defined coordinate system. A and B represents the pan angle and tilt angle of the camera respectively, shown in Fig. 13.12 (b) and (c), and C represents the skew angle, i.e. the orientation of the image plane w.r.t. the camera coordinate system which is nil in our case. From the above relation we can easily derive the perspective matrix (T) of the camera, which is denoted by (13.17)

$$
T = \begin{pmatrix} t_{11} & t_{12} & t_{13} & t_{14} \\ t_{21} & t_{22} & t_{23} & t_{24} \\ t_{31} & t_{32} & t_{33} & t_{34} \end{pmatrix}
$$

$$
= \begin{pmatrix} 1 & 0 & 0 & 0 \\ 0 & 1 & 0 & 0 \\ 0 & 0 & -1/f & 1 \end{pmatrix} \begin{pmatrix} 1 & 0 & 0 & 0 \\ 0 & 1 & 0 & 0 \\ 0 & 0 & -1 & f \\ 0 & 0 & 0 & 1 \end{pmatrix} \begin{pmatrix} 1 & 0 & 0 & 0 \\ 0 & \cos A & -\sin A & 0 \\ 0 & \sin A & \cos A & 1 \\ 0 & 0 & 0 & 1 \end{pmatrix}
$$

$$
\begin{pmatrix} \cos B & 0 & \sin B & 0 \\ 0 & 1 & 0 & 0 \\ -\sin B & 0 & \cos B & 0 \\ 0 & 0 & 0 & 1 \end{pmatrix} \begin{pmatrix} \cos C & -\sin C & 0 & 0 \\ \sin C & \cos C & 0 & 0 \\ 0 & 0 & 1 & 0 \\ 0 & 0 & 0 & 1 \end{pmatrix} \begin{pmatrix} 1 & 0 & 0 & -x_0 \\ 0 & 1 & 0 & -y_0 \\ 0 & 0 & 1 & -z_0 \\ 0 & 0 & 0 & 1 \end{pmatrix} \tag{13.17}
$$

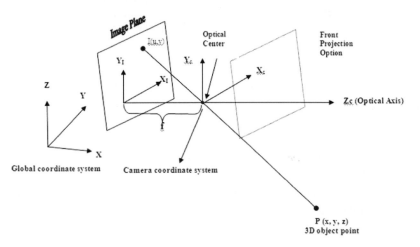

Fig. 13.12 (a) Generalized camera model of 3D projection

(b)

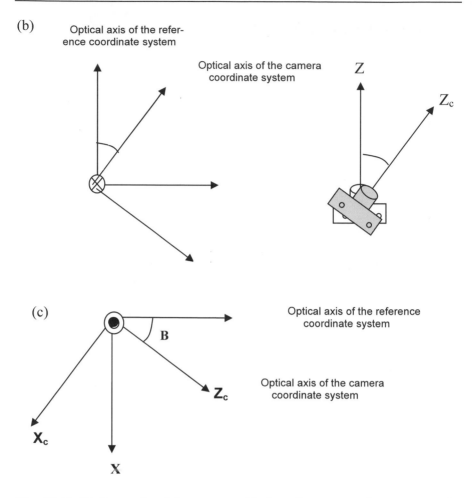

Fig. 13.12 (b) Pan angle of the camera with the reference coordinate system; **(c)** Tilt angle of the camera with the reference coordinate system

13.7 Summary

This chapter briefly highlights robot perception and the process of image formation using perspective projection geometry. The expression for determining the perspective matrix of the camera, which transforms 3D object points onto the image plane, has been derived.

14 Image Capture Program

14.1 Introduction

In the last chapter we saw that vision is one of the important human senses and this is also the case for robots. In this chapter we will discuss how to capture the gray and color images using a camera from the surroundings. Here the image is captured by the camera fitted to the robot and controlled from a remote client. The program is developed in a client–server paradigm. During the run session, the color images are continuously being sent to the client. When the user at the client requests a color image, the color image displayed on the client screen is selected, and when the grayscale image is required the color image is converted to grayscale in the client. As the color images have larger bandwidth, transmitting them directly would require a substantial amount of time and memory as well. Therefore, run-time encoding is used in the server program, which is discussed below.

14.2 Algorithm for Image Capture

The run-length encoding and server and client algorithm is given below. The server program and the client program are given in Listings 14.1 and 14.2 respectively at the website of the book. The program log session is given below and the output is illustrated in Figs. 14.1 and 14.2.

Run Length Encoding Algorithm

Step 1: The color and grayscale images are obtained.
Step 2: $C(x)$ – color pixel at location x.
$\quad\quad$ $G(x)$ – grayscale pixel at location x.
$\quad\quad$ Count = 0, x = 0, x_cur = 0;
Step 3: x_cur = x_cur + 1

If x_cur > Max
Goto step 5.
Step 4: If Absolute(G(x_cur) – G(x)) > range
Count ++
x++
Goto Step 3.
Else
The RGB values of location "x" and count are sent from the server to the client.
Count =0, x = x_cur;
Goto Step 3.
Step 5: The RGB values of location "x" and count are sent from the server to the client.
Step 6: Continue with next image, i.e. Goto Step 1.

Server Program Algorithm

Step 1: Open the server socket and wait for the client to join.
Step 2: Once the client has joined start the run-length algorithm.
Step 3: When the user interrupts, close the sockets and stop the program.

Client Program Algorithm

Step 1: Connect to the server socket if opened.
x = 0, max = 320 × 240 (size of the image)
flag = true, (for color image)
Image() = image pixel array.
Step 2: Keep reading the socket till four integer values are received (RGB and count) from the socket.
Step 3: for I = 0 , I < count , I ++
If flag = true
Image(x)(0) = R,
Image(x)(1) = G,
Image(x)(2) = B,
x++;
Else
Image(x) = 0.11*R+0.56*G+0.33*B
Step 4: If x = max
Display the image.
Else goto Step 2.
Step 5: On user interruption
If color button

flag = true
If gray scale
flag =false
If cross
Goto Step 6.
Step 6: Stop.

Program output log

```
IEEE 1394 interface open request
1 card(s) found, 2 node(s)
Checking card 0, node 0
Vitana api addr: 78080600
Vendor length is 10
Vendor is: VITANA
Model length is 10
Model is: PixeLINK(tm)
Camera found at node 0 0: VITANA PixeLINK(tm)
Camera ISO bandwidth needed: A10
Max_Image_Size_Inq: 05080408
Unit_Size_Inq: 00080008 (0)
Image_Size_Inq: 05000400 (0)
Frame_Rate_min: 0000000E (0)
Frame_Rate_max: 000000A0 (0)
Frame_Rate_def: 00000008 (0)
Flags: 00000000 (0)
PCS2112 ver 0x30
Imager reset starting...
Imager reset succeeded
Imager ready.
Camera ISO speed set to 400 Mb/sec
Camera ISO parameters: 2000000
Opened frame grabber.
Size: 320/1280 240/960
Opened the server port
Client has connected
Size: 320/1280 240/960
Starting DMA
Camera ISO bandwidth needed: A10 (2576)
w, h: 640, 480
frame bytes: 614400
Size: 196608
buf size: 196608
Iso Thread running..., dmafd = 4 (431D7000 42A16010)
DBS: 161  FN: 1  QPC: 1 MAX_DBS: 4
```

```
DBC count error... 108 142 3 150084
DBC count error... 12 42 2 5148
DBC count error... 52 62 4 83272
DBC count error... 228 244 3 47500
DBC count error... 35 69 3 120996
DBC count error... 91 129 3 106072
DBC count error... 104 130 4 124016
DBC count error... 138 175 2 116968
DBC count error... 235 17 3 121468
DBC count error... 164 195 3 137528
DBC count error... 87 121 3 115876
DBC count error... 234 0 3 57524
DBC count error... 83 118 3 36856
DBC count error... 178 206 3 156204
DBC count error... 47 66 3 82984
DBC count error... 73 107 3 112568
DBC count error... 18 46 3 158060
DBC count error... 51 86 3 90124
DBC count error... 117 152 3 137812
DBC count error... 32 66 2 161312
DBC count error... 192 226 3 118320
Closing video device and TCP connection ...
Stopping DMA
Aria: Received signal 'SIGINT'. Shutting down.
Closing video device and TCP connection ...
Stopping DMA
ISO thread terminated
```

Fig. 14.1. Color image from image capture program

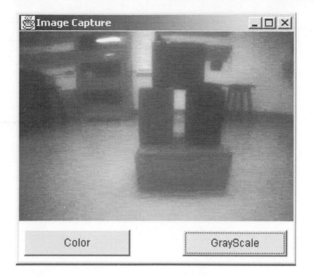

Fig. 14.2. Grayscale image from image capture program

14.3 Summary

It can be seen in Fig. 14.1 that there are two buttons "Color" and "Gray-Scale". When the "Color" button is pressed, the color image from the robot server is displayed in the Java frame, while when the "GrayScale" button is pressed the color image is converted into grayscale and displayed. Generally the image manipulations are carried out on grayscale images.

15 Building 3D Perception Using a Kalman Filter

15.1 Introduction

A Kalman filter is a recursive digital filter [Brown, 1997] that acts as a set of incoming data structures to estimate the parameters of a system. Ayache employed Kalman filtering [Ayache, 1987; Ayache, 1991] for 3D reconstruction of images. In fact the Kalman filter can be used to construct 2D lines from noisy 2D image points, affine 3D points from 2D image points, affine 3D lines from noisy 2D image points or from affine 2D lines or from 3D points, and 3D planes either from 3D points or 3D lines. Before employing Kalman filtering for 3D reconstruction, we will briefly outline the minimal parametric representation of 2D lines, 3D lines, and 3D planes. After the minimal representation, these parameters can be directly used to recursively update the filter equation in order to find the estimators of the system. A 3D reconstruction is required, generally, to find the depth information of an object. The images by which the depth can be measured are usually called stereo images. A number of cameras are employed to extract the features of the stereo images. The number of cameras is generally restricted to three for most image processing applications. The significance of the Kalman filter in 3D reconstruction lies in streamlining the process of feature extraction through multiple cameras. In this chapter, we present some experiments to construct (a) 3D points from noisy 2D image points, (b) a 3D line from 3D points and (c) a 3D plane from 3D points. Let us first discuss the possible minimal representation of 2D lines, 3D lines, and 3D planes.

15.2 Minimal Representation

A 2D line AB can be minimally best represented by two parameters $\mathbf{a}$ and $\mathbf{p}$ as evident from Fig. 15.1. The advantage of this parameterization is that the equation of the lines is linear in the parameters (a, p), which is essential in the formulation of the recursive Kalman filtering equation. Secondly the state vector which is derived from these parameters satisfies the inequality check criteria of the recursive Kalman filter.

Similarly, the 3D line CD can be minimally represented by four parameters **a**, **b**, **p**, **q**, as shown in Fig. 15.2 and the plane EFGH can be represented by three parameters **a**, **b**, **p** as shown in Fig. 15.3.

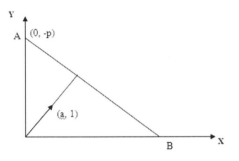

$a\,x + y + p = 0$ (when the line is not parallel to the Y-axis), or
$x + a\,y + p = 0$ (when the lines are not parallel to the X-axis)

Fig. 15.1. A 2D line AB that passes through $(0,-p)$, normal to the line, passing through $(0,0)$ and $(a,1)$, can be represented by two parameters **a** and **p**

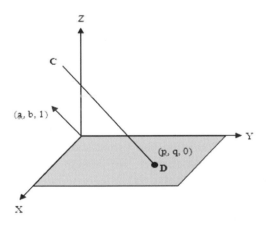

$x = a\,z + p$ and $y = b\,z + q$ (when the line is not orthogonal to the Z-axis)
$y = a\,x + p$ and $z = b\,x + q$ (when the line is not orthogonal to the X-axis)
$z = a\,y + p$ and $x = b\,y + q$ (when the line is not orthogonal to the Y-axis)

Fig. 15.2. A 3D affine line CD that passes through the XY plane at a point $(p, q, 0)$ and having the direction vector $(a, b, 1)^{\mathrm{T}}$ can be represented by four parameters **a, b, p, q**

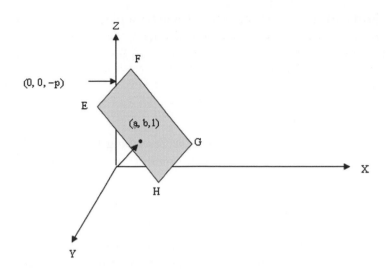

$a\,x + b\,y + z + p = 0$ (when the planes are not parallel to the Z-axis)
$x + a\,y + b\,z + p = 0$ (when the planes are not parallel to the X-axis)
$b\,x + y + a\,z + p = 0$ (when the planes are not parallel to the Y-axis)

Fig. 15.3. A 3D affine plane EFGH that passes through $(0,0,-p)$ and normal to the plane passing through $(0, 0, 0)$ and $(a, b, 1)$ can be represented by three parameters **a, b, p**

The parameters mentioned above help in representing the affine lines and planes in a minimum, complete and unambiguous manner, which can be used in subsequent estimation. Further, it is to be noted that the representations also satisfy the differentiability criteria to allow linearization of the measurement equation in the formulation of the recursive Kalman filter, to be covered in the next section.

15.3 Recursive Kalman Filter

A Kalman filter is a digital filter that attempts to minimize the measurement noise from estimating the unknown parameters, linearly related with a set of measurement variables. The most important characteristic of this filter is that it allows a recursive formulation and the user can improve the accuracy of the estimation to a desired level at the cost of new measurement inputs.

Let

x_i be a measurement vector of dimension ($m_i \times 1$),

$\mathbf{K}_i$ be the filter gain matrix of dimension ($n \times p_i$),
$\mathbf{a}_i$ be the estimator vector of dimension ($n \times 1$),
$\mathbf{M}_i$ be a system matrix of dimension ($p_i \times n$) such that

$$M_i = \frac{\partial f_i}{\partial a} \tag{15.1}$$

where, f_i ($\mathbf{x}_i$, $\mathbf{a}$) = 0 is a set of equations describing the relationships between a parameter vector a and the measurement variable vector x_i,

$$y_i = -f_i(x_i^*, a_{i-1}) + [\frac{\partial f_i}{\partial a}]_{(x_i^*, a_{i-1})}(a - a_{i-1}^*)$$

is a modified measurement vector of dimension (p_i x 1), obtained by linearization of f_i ($\mathbf{x}_i$, $\mathbf{a}$) = 0 around $\mathbf{x_i} = \mathbf{x_{i-1}}$ and $\mathbf{a} = \mathbf{a}^*$ using Taylor series.

$$W_i = E[w_i w_i^T] = [\frac{\partial f_i}{\partial x}]_{(x_i^*, a_{i-1})} \wedge_i [[\frac{\partial f_i}{\partial x}]_{(x_i^*, a_{i-1})}]^T \tag{15.2}$$

where $w_i = (\partial f_i / \partial x) (x_i - x_i^*)$ is the measurement noise vector of dimension ($p_i \times 1$) and Λ_i is a positive symmetric matrix.

$S_i = E[(a_i - a_{i*})(a_i - a_i^*)^T]$ is the error covariance matrix of the estimator $\mathbf{a}$.

The recursive formulation of an EKF includes the following three steps.

$$K_i = S_{i-1} M_i^{\ T} (W_i + M_i S_{i-1} M_i^{\ T})^{-1} \tag{15.3}$$
$$a_i^* = a_{i-1}^* + K_i (y_i - M_i a_{i-1}^*) \tag{15.4}$$
$$S_i = (I - K_i M_i) S_{i-1} \tag{15.5}$$

The algorithm is initialized with a large S_0. The values of y_i, M_i, W_i are computed following their above definitions. a_0 is initialized as a null vector. The algorithm then continues iterating in sequence until S_i comes below a predefined threshold. The resulting a_i after termination of the algorithm is the desired estimator. A schematic diagram depicting the use of EKF in estimating noise-free geometric parameters from noisy 2D images is presented in Fig. 15.4.

Noise State Estimation

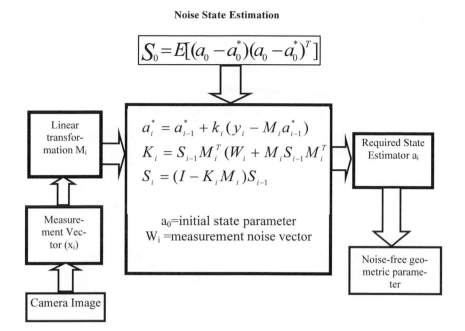

$$S_0 = E[(a_0 - a_0^*)(a_0 - a_0^*)^T]$$

Linear transformation M_i	$a_i^* = a_{i-1}^* + k_i(y_i - M_i a_{i-1}^*)$ $K_i = S_{i-1} M_i^T (W_i + M_i S_{i-1} M_i^T$ $S_i = (I - K_i M_i) S_{i-1}$	Required State Estimator a_i

Measurement Vector (x_i)

a_0=initial state parameter
W_i =measurement noise vector

Camera Image

Noise-free geometric parameter

Fig. 15.4. Schematic diagram of a recursive Kalman filter

15.4 Experiments and Estimation

The experimental set-up was designed for image abstraction of a wooden block from different angles by a mobile robot for 3D reconstruction. Fig. 15.5 illustrates the trajectory of the camera movement by the robot around the block W. Six snaps have been taken in each of the four segments separated by the lines aa' and bb', bb' and cc', cc' and dd', dd' and aa' respectively. These four sets of images are illustrated by Set I, Set II, Set III and Set IV, respectively. To illustrate the use of Kalman filter, we first explore the possibility of reconstructing the 3D points from multiple 2D points, followed by a 3D line from 3D points, and a 3D plane from 3D points. The detailed program is covered in the next chapter.

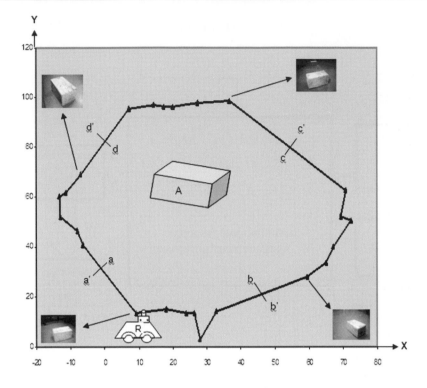

Fig. 15.5. The trajectory of camera movement by a robot R around the block A to grab its image from 24 different locations, denoted by triangles. Grabbed images at a few locations are shown

Set I

Visual Map_1
Pinhole Position: $x=9$, $y=14$, $z=28.5$,
$A = 130°$, $B=90°$

Visual Map_2
Pinhole Position: $x=18$, $y=15.5$, $z=28.5$,
$A=0°$, $B=90°$

Visual Map_3
Pinhole Position: $x=24$, $y=13.5$, $z=28.5$,
$A=0°$ $B=90°$

Visual Map_4
Pinhole Position: $x=26.5$, $y=13.7$, $z=28.5$,
$A=13°$ $B=90°$

Visual Map_5
Pinhole Position: $x=28$, $y=3$, $z=28.5$,
$A=0°$ $B=90°$

Visual Map_6
Pinhole Position: $x=32.5$, $y=14.8$, $z=28.5$,
$A=15°$ $B=90°$

Set II

Visual Map_1
Pinhole Position: x=59.5, y=28.2, z=28.5,
A= 53°, B=130°

Visual Map_2
Pinhole Position: x=65, y=34, z=28.5,
A= 45°, B=90°

Visual Map_3
Pinhole Position: x=67.0, y=40.0, z=28.5,
A= 71°, B=90°

Visual Map_4
Pinhole Position: x=72.3, y=50.1, z=28.5,
A= 78°, B=90°

Visual Map_5
Pinhole Position: x=69.5, y=52.0, z=28.5,
A= 90°, B=90°

Visual Map_6
Pinhole Position: x=70.7, y=62.5, z=28.5,
A= 100°, B=90°

Set III

Visual Map_1
Pinhole Position: x=36.5, y=98.5, z=28.5,
A= 160°, B=90°

Visual Map_2
Pinhole Position: x=27.5, y=97.5, z=28.5,
A= 180°, B=90°

Visual Map_3
Pinhole Position: x=20.0, y=96.0, z=28.5,
A= 185°, B=90°

Visual Map_4
Pinhole Position: x=17.5, y=96.0, z=28.5,
A= 190°, B=90°

Visual Map_5
Pinhole Position: x=14.5, y=96.5, z=28.5,
A= 192°, B=90°

Visual Map_6
Pinhole Position: x=7.0, y=95.0, z=28.5,
A= 200°, B=90°

Set IV

Visual Map_1
Pinhole Position: x=−7.0, y=69.0, z=27.0,
A= 242°, B=140°

Visual Map_2
Pinhole Position: x=−11.5, y=62.0, z=27.0,
A= 245°, B=140°

Visual Map_3
Pinhole Position: x=−13.5, y=59.7, z=28.5,
A=270°, B=90°

Visual Map_4
Pinhole Position: x=−13.0, y=52.0, z=28.5,
A= −80°, B=90°

Visual Map_5
Pinhole Position: x= −8.0, y=46.5, z=28.5,
A=−70°, B=90°

Visual Map_6
Pinhole Position: x=−6.5, y=40.8, z=28.5,
A=−65°, B=90°

15.4.1 Reconstruction of 3D Points

The coordinates of the 2D vertices of the block W, measured from these six images, have been used recursively as the input to a Kalman filter for the reconstruction of their 3D coordinates. Fig. 15.6 illustrates the input and output parameters of a Kalman filter employed for reconstruction of 3D coordinates of a vertex A from six 2D image coordinates A_1 through A_6 of the same vertex A. The equations used for 3D reconstruction in the present context are given by the following expressions:

$$y_i = \begin{bmatrix} t_{34} \cdot u_i - t_{34} \\ t_{34} \cdot v_i - t_{34} \end{bmatrix} \tag{15.6}$$

$$M_i = \begin{bmatrix} -(u_i t_3 - t_1)^T \\ -(v_i t_3 - t_2)^T \end{bmatrix} \tag{15.7}$$

where M_i is a 2×3 matrix and y_i is a two dimensional vector, t_{ij} means the i-th row and j-th column element of the perspective matrix T (discussed in the last chapter) and t_i is the first three elements of the i-th row of the same matrix T. (u_i, v_i) is the 2D image points of the vertex.

The measurement error matrix w_i is a 2×2 matrix estimated by the following expression:

$$w_i = \frac{\partial f}{\partial x_i} = \begin{pmatrix} -t_3 \cdot a_{i-1} - t_{34} & 0 \\ 0 & -t_3 \cdot a_{i-1} - t_{34} \end{pmatrix} \tag{15.8}$$

Further, $\mathbf{a}$ = estimation vector = $[x, y, z]^T$ with initial value of $[0, 0, 0]^T$; and with a very large initial covariance matrix S_o, which is of size 3×3 in this case.

The first two iterations of the estimation process of reconstructing the 3D point from the multiple 2D image points is given in Listings 15.1 and 15.2 respectively. The inputs of Box 1 are a 2D point $(u_1, v_1) = (-2.7, 1.3)$ and a set of camera parameters $x_0=36.5$, $y_0=98.5$, $z_0=28.5$, $A=2.793$, $B=-1.97$, $C=0$. The output of Box 1 is the 3D point $(x, y, z) = (22.11, 6.84, -19.57)$ and the covariance matrix is

$$\begin{bmatrix} 232.6 & 1192.9 & 627.5 \\ 1192.9 & 7699.3 & 4021.5 \\ 627.5 & 4021.5 & 2161.7 \end{bmatrix}$$

Similarly the input of Box 2 is the 2D point $(u_2, v_2)= (-2.25, 1.3)$ and a new set of camera parameters x_0=27.5, y_0=97.5, z_0=28.5, A=3.14, B=-1.97, C=0. The output of Box 2 is the 3D point (x, y, z)= (32.04, 63.64, 10.44) and the covariance matrix is

$$\begin{bmatrix} 22.76 & -8.92 & -4.42 \\ -8.92 & 805.88 & 413.28 \\ -4.43 & 413.28 & 240.84 \end{bmatrix}$$

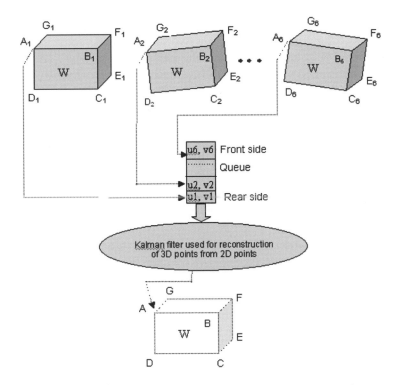

Fig. 15.6. Extraction of the 3D points from six 2D image points. 2D points A_1 through A_6 have been fused to point A by Kalman filtering. The other points B, C, D, E, F, F in the figure have been reconstructed similarly

Listing 15.1

First iteration in the estimation process of 3D point reconstruction from noisy 2D points using a Kalman filter.

Initial State vector $a = [a1_0, a2_0, a3_0] = [0\ 0\ 0]^T$ (initial assumption).
Initial error covariance matrix S_0 (a very large value) =

$$\begin{pmatrix} 9999 & 0 & 0 \\ 0 & 9999 & 0 \\ 0 & 0 & 9999 \end{pmatrix}$$

Substituting the first value of the measurement vector (x_i) that is taken from image $= (u_1, v_1) = (-2.7, 1.3)$
The camera parameters: $(x_0, y_0, z_0, A, B, C) = (36.5, 98.5, 28.5, 2.793, -1.97, 0)$
The perspective matrix for the above camera parameter can be calculated by using following expression.

$$T = \begin{pmatrix} t_{11} & t_{12} & t_{13} & t_{14} \\ t_{21} & t_{22} & t_{23} & t_{24} \\ t_{31} & t_{32} & t_{33} & t_{34} \end{pmatrix} = \begin{pmatrix} -0.135 & 0.049 & 0 & 0.09 \\ 0.019 & 0.053 & -0.13 & -2.16 \\ -0.002 & -0.008 & 1.0 & 1.0 \end{pmatrix}$$

$$y_1 = \begin{pmatrix} t_{34}*u_1 - t_{14} \\ t_{34}*v_1 - t_{24} \end{pmatrix} = \begin{pmatrix} -2.79 \\ 3.45 \end{pmatrix}$$

$$M_1 = \begin{pmatrix} -(u_1*[t_{31}\ t_{32}\ t_{33}] - [t_{11}\ t_{12}\ t_{13}]) \\ -(v_1*[t_{31}\ t_{32}\ t_{33}] - [t_{21}\ t_{22}\ t_{23}]) \end{pmatrix} = \begin{pmatrix} -0.143 & 0.027 & -0.009 \\ 0.023 & 0.063 & -0.128 \end{pmatrix}$$

$$W_1 = \begin{pmatrix} -([t_{31}\ t_{32}\ t_{33}]*[a1_0\ a2_0\ a3_0]-t_{34}) & 0 \\ 0 & -([t_{31}\ t_{32}\ t_{33}]*[a1_0\ a2_0\ a3_0]-t_{34}) \end{pmatrix} = \begin{pmatrix} 1.0 & 0 \\ 0 & 1.0 \end{pmatrix}$$

$$K_1 = \text{Kalman Gain} = S_0*M_1^T*(W_1 + M_1*S_0*M_1^T) = \begin{pmatrix} -6.67 & 1.01 \\ 1.33 & 3.05 \\ -0.54 & 6.01 \end{pmatrix}$$

$$\text{State estimator } a_1 = a_0 + K_1*(y_1 - M_1*a_0) = \begin{pmatrix} 22.11 \\ 6.84 \\ -19.57 \end{pmatrix}$$

$$\text{Error covariance matrix } S_1 = (I - K_1*M_1)*S_0 = \begin{pmatrix} 232.6 & 1192.9 & 627.5 \\ 1192.9 & 7699.3 & 4021.5 \\ 627.5 & 4021.5 & 2161.7 \end{pmatrix}$$

Listing 15.2

Second iteration in the estimation process of 3D point reconstruction from noisy 2D points using a Kalman filter.

Initial State vector $\mathbf{a_1} = [a1_1, a2_1, a3_1] = [22.11\ 6.84 -19.57\ ]^T$ (from 1st iteration).
Initial error covariance matrix S_1 (from 1st iteration)=

$$\begin{pmatrix} 232.6 & 1192.9 & 627.5 \\ 1192.9 & 7699.3 & 4021.5 \\ 627.5 & 4021.5 & 161.7 \end{pmatrix}$$

Substituting the second measurement vector: $(u_2, v_2) = (-2.25, 1.3)$
The camera parameters for the second iteration: $(x_0, y_0, z_0, A, B, C) = (27.5, 97.5, 28.5, 3.14, -1.97, 0\ )$
The perspective matrix T_2 for the above camera parameter is

$$\begin{pmatrix} -0.154 & 0 & 0 & 4.226 \\ 0 & 0.060 & -0.141 & -1.849 \\ 0 & -0.009 & 1.0 & 1.0 \end{pmatrix}$$

$$y_2 = \begin{bmatrix} t_{34}*u_2 - t_{14} \\ t_{34}*v_2 - t_{24} \end{bmatrix} = \begin{bmatrix} -6.48 \\ 3.15 \end{bmatrix}$$

$$M_2 = \begin{bmatrix} -(u_2*[t_{31}\ t_{32}\ t_{33}] - [t_{11}\ t_{12}\ t_{13}]) \\ -(v_2*[t_{31}\ t_{32}\ t_{33}] - [t_{21}\ t_{22}\ t_{23}]) \end{bmatrix} = \begin{bmatrix} -0.154 & -0.020 & -0.009 \\ 0 & 0.072 & -0.136 \end{bmatrix}$$

$$W_2 = \begin{bmatrix} -([t_{31}\ t_{32}\ t_{33}]*[a1_1\ a2_1\ a3_1] -t_{34}) & 0 \\ 0 & -([t_{31}\ t_{32}\ t_{33}]*[\ a1_1\ a2_1\ a3_1] -t_{34}) \end{bmatrix} = \begin{bmatrix} 1.027 & 0 \\ 0 & 1.027 \end{bmatrix}$$

$$K_2 = \text{Kalman Gain} = S_1*M_2^T*(W_2 + M_2*S_1*M_2^T) = \begin{pmatrix} -3.192 & -0.039 \\ -18.276 & 1.756 \\ -9.640 & -2.932 \end{pmatrix}$$

$$\text{State estimator } \mathbf{a_2} = a_1 + K_2*(y_2 - M_2*a_1) = \begin{pmatrix} 32.04 \\ 63.64 \\ 10.44 \end{pmatrix}$$

$$\text{Next Error covariance matrix } S_2 = (I - K_2*M_2)*S_1 = \begin{pmatrix} 22.79 & -8.92 & -4.42 \\ -8.92 & 805.88 & 413.28 \\ -4.43 & 413.28 & 240.84 \end{pmatrix}$$

and this will continue until all the measurement vector is exhausted or up to the specified covariance error.

The response of the Kalman filter with subsequent 2D image point inputs for reconstruction of the 3D points is presented in figure 15.6. The 3D reconstruction of vertex A of the wooden block W is estimated recursively by providing the 2D points as given below.

1. $(u_1, v_1) = (-2.7, 1.3)$ $(x, y, z) = (22.11, 6.84, -19.57)$
2. $(u_2, v_2) = (-2.25, 1.3)$ $(x, y, z) = (32.04, 63.64, 10.44)$
3. $(u_3, v_3) = (-3.0, 1.5)$ $(x, y, z) = (31.66, 59.26, 8.0)$
4. $(u_4, v_4) = (-2.9, 1.0)$ $(x, y, z) = (31.47, 60.53, 9.05)$
5. $(u_5, v_5) = (-4.0, 0.9)$ $(x, y, z) = (30.48, 65.11, 11.40)$
6. $(u_6, v_6) = (-3.7, 0.7)$ $(x, y, z) = (31.82, 60.42, 8.86)$

The recursive estimation of the 3D points is shown graphically in Fig. 15.7, where it is clear that the accuracy in the 3D reconstruction increases with the number of 2D image points. Fig. 15.8 shows the covariance error versus the number of iterations.

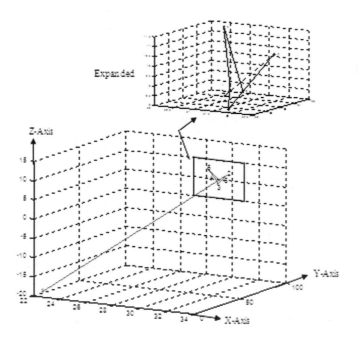

Fig. 15.7. The 3D reconstruction points with incoming stream of 2D points A1–A6

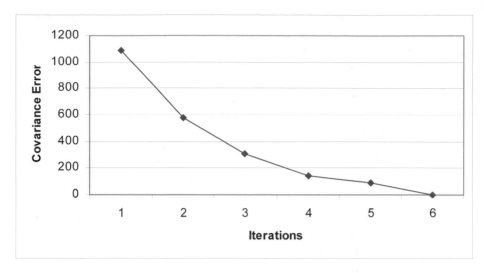

Fig. 15.8. Improvements in the accuracy of the 3D reconstruction with increase in the number of 2D image points

15.4.2 Reconstruction of a 3D Line

For reconstruction of a 3D lines from 3D points, Kalman filtering can be employed. The estimator, measurement parameters and other components of the filter equation are represented by the following expressions.

$$y_i = \begin{bmatrix} x_i \\ y_i \end{bmatrix} \qquad (15.9)$$

$$M_i = \begin{bmatrix} z_i & 0 & 1 & 0 \\ 0 & z_i & 0 & 1 \end{bmatrix} \qquad (15.10)$$

where M_i is a 2×4 matrix and y_i is a two-dimensional vector, and (x_i, y_i, z_i) are the coordinates of the 3D points from multiple sources serving as the measurement vector.

The measurement error matrix w_i (2×3) can be estimated by the following expression.

$$\frac{\partial f}{\partial x_i} = \begin{pmatrix} -1 & 0 & a_{i-1} \\ 0 & -1 & b_{i-1} \end{pmatrix} \tag{15.11}$$

Further, $\mathbf{a_0}$ is the initial estimation vector $= [a_0, b_0, p_0, q_0]^T$ which may be assumed any arbitrary value such as $[0\ 0\ 0\ 0]$ or may be computed from first two points as follows:

$$a_0 = \frac{x_2 - x_1}{z_2 - z_1}; \qquad\qquad b_0 = \frac{y_2 - y_1}{z_2 - z_1}$$

$$p_0 = \frac{z_2 x_1 - z_1 x_2}{z_2 - z_1}; \qquad q_0 = \frac{z_2 y_1 - z_1 y_2}{z_2 - z_1} \tag{15.12}$$

where (x, y, z) denotes the coordinate of the input 3D points and $a, b, p,$ and q describe the parameters of the reconstructed 3D lines in the above expression.

The first two iterations in the process of estimation of the 3D line reconstruction from 3D points is given in Listings 15.3 and 15.4 respectively. The input of Box 3 is a 3D point $(x, y, z) = (8.5, 2.0, 2.5)$ along with the initial state estimator $\mathbf{a_0}$ and covariance matrix S_0. The output of Listing 15.3 is the 3D line parameters $(a, b, p, q) = (3.5, 0.79, 0.13, 0.03)$ and the covariance matrix S_1. Similarly the input of Listing 15.4 is the 3D point $(x, y, z) = (9.0, 2.5, 2.7)$ along with S_1 and the output derived is the updated 3D line parameters $(a, b, p, q) = (2.8, 2.20, 1.49, -3.43)$ with covariance matrix S_2.

Listing 15.3

First iteration in the process of 3D line estimation from 3D points using a Kalman filter.

Initial state vector $\mathbf{a} = [a_0, b_0, p_0\ q_0] = [0\ 0\ 0\ 0]^T$ (initial assumption).
Initial error covariance matrix S_0 (a very large value) =

$$
\begin{pmatrix}
9999 & 0 & 0 & 0 \\
0 & 9999 & 0 & 0 \\
0 & 0 & 999 & 0 \\
0 & 0 & 0 & 999
\end{pmatrix}
$$

Submitting the first measurement vector $(x_1, y_1, z_1) = (8.5\ 2.0\ 2.5)$

$$
y_1 = \begin{pmatrix} x_1 \\ y_1 \end{pmatrix} = \begin{pmatrix} 8.5 \\ 2.0 \end{pmatrix}
$$

$$
M_1 = \begin{pmatrix} z_1 & 0 & 1 & 0 \\ 0 & z_1 & 0 & 1 \end{pmatrix} = \begin{pmatrix} 2.5 & 0 & 1 & 0 \\ 0 & 2 & 0 & 1 \end{pmatrix}
$$

$$
W_1 = \begin{pmatrix} -1 & 0 & a_0 \\ 0 & -1 & b_0 \end{pmatrix} \begin{pmatrix} -1 & 0 \\ 0 & -1 \\ a_0 & b_0 \end{pmatrix} = \begin{pmatrix} -1.0 & 0 \\ 0 & 1.0 \end{pmatrix}
$$

$$
K_1 = \text{Kalman Gain} = S_0 * M_1^T * (W_1 + M_1 * S_0 * M_1^T) = \begin{pmatrix} 0.394 & 0 \\ 0 & 0.394 \\ 0.157 & 0 \\ 0 & 0.016 \end{pmatrix}
$$

State estimator $\mathbf{a_1} = a_0 + K_1 * (y_1 - M_1 * a_0) = [\ a_1, b_1, p_1, q_1]^T$

$$
= [3.34\ 0.78\ 0.13\ 0.031]^T
$$

Next Error covariance matrix $S_1 = (I - K_1 * M_1) * S_0 =$

$$
\begin{pmatrix}
157.17 & 0 & -393.32 & 0 \\
0 & 157.48 & 0 & -393.30 \\
-393.32 & 0 & 983.28 & 0 \\
0 & -393.30 & 0 & 983.28
\end{pmatrix}
$$

Listing 15.4

Second Iteration in the process of 3D line estimation from 3D points using a Kalman filter.

Initial State vector $\mathbf{a} = [a_0, b_0, p_0\ q_0] = [3.34\ 0.78\ 0.13\ 0.031]^T$ (from 1st iteration).
Initial error covariance matrix S_1 (from 1st iteration) =

$$
\begin{pmatrix}
157.17 & 0 & -393.32 & 0 \\
0 & 157.48 & 0 & -393.30 \\
-393.32 & 0 & 983.28 & 0 \\
0 & -393.30 & 0 & 983.28
\end{pmatrix}
$$

Now substituting the second measurement vector $(x_2, y_2, z_2) = (9.0, 2.5, 2.7)$

$$
y_2 = \begin{bmatrix} x_2 \\ y_2 \end{bmatrix} = \begin{bmatrix} 9.0 \\ 2.5 \end{bmatrix}
$$

$$
M_2 = \begin{bmatrix} z_2 & 0 & 1 & 0 \\ 0 & z_2 & 0 & 1 \end{bmatrix} = \begin{bmatrix} 2.7 & 0 & 1 & 0 \\ 0 & 2.7 & 0 & 1 \end{bmatrix}
$$

$$
W_1 = \begin{bmatrix} -1 & 0 & a_1 \\ 0 & -1 & b_1 \end{bmatrix} \begin{bmatrix} -1 & 0 \\ 0 & -1 \\ a_1 & b_1 \end{bmatrix} = \begin{pmatrix} 10.12 & 2.62 \\ 2.64 & 1.62 \end{pmatrix}
$$

$$
K_2 = \text{Kalman Gain} = S_1 * M_2{}^T * (W_2 + M_2 * S_1 * M_2{}^T) = \begin{pmatrix} 2.12 & -0.64 \\ -0.666 & 3.76 \\ -5.38 & 1.64 \\ 1.64 & -9.28 \end{pmatrix}
$$

State estimator $\mathbf{a}_2 = a_1 + K_2 * (y_2 - M_2 * a_1) = [a_2, b_2, p_2, q_2]^T$
$$= [2.81 \quad 2.20 \quad 1.49 \quad -3.44\ ]^T$$

Next Error covariance matrix $S_2 = (I - K_2 * M_2) * S_1 =$

$$
\begin{pmatrix}
93.59 & 19.93 & -232.08 & -49.14 \\
19.94 & 41.54 & -50.60 & -107.51 \\
-232.07 & -50.60 & 574.39 & 124.63 \\
-49.14 & -107.51 & 124.63 & 278.81
\end{pmatrix}
$$

and this will continue until all the measurement vector is exhausted or up to the specified covariance error.

15.4.3 Reconstruction of a 3D Plane

For the reconstruction of the 3D plane, the Kalman filter can also be employed with a different set of measurement and estimation vectors. For the case when the plane is not parallel to Z axis, the vectors can be represented as follows:

a = estimation vector of the plane = $[a, b, p]$

where a, b, p have the usual meaning indicated in Fig. 15.3. The measurement vector (x_i) is the 3D points (x, y, z)
    We can choose the measurement equation as:

$$f_i(x, a) = ax_i + by_i + z_i + p = 0 \tag{15.13}$$

After linearization, the following expressions can be derived for the estimation.

$$y_i = -z_i \tag{15.14}$$

$$M_i = \begin{bmatrix} x_i & y_i & 1 \end{bmatrix} \tag{15.15}$$

$$\frac{\partial f}{\partial x} = \begin{bmatrix} a_{i-1} & b_{i-1} & 1 \end{bmatrix} \tag{15.16}$$

The first two iterations of the estimation towards the construction of a 3D plane from the 3D points is given in Listing 15.5 and 15.6 respectively. The input of Box 5 is a 3D point $(x, y, z) = (0, 0, 3)$ along with an initial estimator a_0 and covariance matrix S_0. The output of Listing 15.5 is the 3D plane parameters $(a, b, p) = (0, 0, -2.997)$. Similarly the input of Listing 15.6 is the 3D point $(x, y, z) = (3.2, 2.2, 3.1)$ along with a_1 and S_1 estimated in Box 5 and the output of box 6 is the 3D plane parameters $(a, b, p) = (-0.022, -0.015, -2.997)$.

Listing 15.5

First iteration in the process of 3D plane reconstruction from 3D points using a Kalman filter.

Initial State vector $\mathbf{a_0} = [a_0, b_0, p_0] = [0\ 0\ 0]^T$ (initial assumption).

Initial error covariance matrix S_0 (a very large value) =

$$\begin{pmatrix} 9999 & 0 & 0 \\ 0 & 9999 & 0 \\ 0 & 0 & 999 \end{pmatrix}$$

Submitting the first measurement vector $[x_1, y_1, z_1] = [0\ 0\ 3.0]$

$y_1 = [-z_1] = [-3.0]$

$M_1 = [x_1 \quad y_1 \quad 1] = [0\ 0\ 1]$

$W_1 = \begin{pmatrix} a_0 & b_0 & 1 \end{pmatrix} \begin{pmatrix} a_0 \\ b_0 \\ 1 \end{pmatrix} = [1]$

$K_1 = $ Kalman Gain $= S_0 * M_1^T * (W_1 + M_1 * S_0 * M_1^T) = \begin{pmatrix} 0 \\ 0 \\ 0.999 \end{pmatrix}$

State estimator $\mathbf{a_1} = a_0 + K_1 * (y_1 - M_1 * a_0) = [a_1, b_1, p_1,]^T = [0 \quad 0 \quad -2.997]^T$

Next Error covariance matrix $S_1 = (I - K_1 * M_1) * S_0 =$

$$\begin{pmatrix} 9999 & 0 & 0 \\ 0 & 9999 & 0 \\ 0 & 0 & 0.998 \end{pmatrix}$$

Listing 15.6

Second iteration in the process of 3D plane reconstruction from 3D points using a Kalman filter.

Initial State vector $\mathbf{a_1} = [a_0, b_0, p_0] = [0 \quad 0 \quad -2.997]^T$ (from 1st iteration).

Initial error covariance matrix S_1 (from 1st iteration) =

$$\begin{pmatrix} 9999 & 0 & 0 \\ 0 & 9999 & 0 \\ 0 & 0 & 0.998 \end{pmatrix}$$

submitting the second measurement vector $[x_2, y_2, z_2] = [3.2\ 2.2\ 3.1]$

$$Y_2 = \quad [-z_2] \quad = \quad [-3.1]$$

$$M_2 = \quad [x_2 \quad y_2 \quad 1] = [3.2 \quad 2.2 \quad 1]$$

$$W_1 = \quad \begin{bmatrix} a_1 & b_1 & 1 \end{bmatrix} \begin{bmatrix} a_1 \\ b_1 \\ 1 \end{bmatrix} = [1]$$

$$K_2 = \text{Kalman Gain} = S_1{}^*M_2{}^{T*}(W_2 + M_2{}^*S_1{}^*M_2{}^T) = \begin{pmatrix} 0.212 \\ 0.145 \\ 0 \end{pmatrix}$$

State estimator $\mathbf{a_2} = a_1 + K_2{}^*(y_2 - M_2{}^*a_1) = [a_2, b_2, p_2,]^T$

$$= [-0.021 \quad 0.015 \quad -2.997]^T$$

Next Error covariance matrix $S_2 = (I - K_2{}^*M_2){}^*S_1 =$

$$\begin{pmatrix} 32.09 & -4667.9 & -0.211 \\ -4667.9 & 6789.8 & -0.145 \\ -0.212 & -0.212 & 0.998 \end{pmatrix}$$

and this will continue until all the measurement vector is exhausted or up to the specified covariance error.

15.5 Correspondence Problem in 3D Recovery

Let us consider two images of the same block W: images W_1 and W_2 shown in Fig. 15.9, which include possibly six and sometimes seven common points from two viewing positions. However, finding the correspondence between the vertices of the two images is difficult. This is usually known as *the correspondence problem* [Dean, 1995]. The correspondence problem can be overcome by estimating the minimal Euclidean distance between the 3D points obtained from two different positions (angles of view). For example, let us consider the two sets of 3D points of the same block W from two distinct angles of view, estimated by the Kalman filter. In order to identify the corresponding vertex between two sets of 3D data points, the minimal Euclidean distance is estimated for each point with all the points of the other set. When the estimated distance is less than a threshold value, the points from both lists are said to be corresponding points and deleted from both lists. The two sets of 3D data points shown in the Fig. 15.9 are estimated from distinct locations.

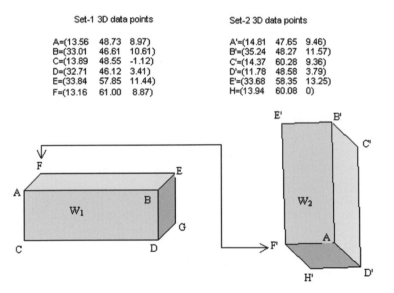

Set-1 3D data points

A=(13.56 48.73 8.97)
B=(33.01 46.61 10.61)
C=(13.89 48.55 -1.12)
D=(32.71 46.12 3.41)
E=(33.84 57.85 11.44)
F=(13.16 61.00 8.87)

Set-2 3D data points

A'=(14.81 47.65 9.46)
B'=(35.24 48.27 11.57)
C'=(14.37 60.28 9.36)
D'=(11.78 48.58 3.79)
E'=(33.68 58.35 13.25)
H=(13.94 60.08 0)

Fig. 15.9. Illustrating the correspondence problem. The correspondence problem aims at mapping point A to A', B to B' and so on after rotating W_1 around the Y and Z axes

Now computing the Euclidean distance between F and A' through H', it is clear that the distance between F and F' is the least. Thus point F' in image W_2 corresponds to F in image W_1.

15.6 Summary

This chapter first presented the minimal representation for the 2D line, 3D line and 3D plane and then discussed the technique for estimation of the 3D parameter of the planar objects using an extended Kalman filter. The results obtained from computer simulations demonstrate that the 3D surface for planar objects can be reconstructed from multiple 2D images of the same object. In the next chapter the client–server program for robot perception will be covered.

16 Program for 3D Perception

16.1 Introduction

A client–server program for 3D perception of a planar object using a Kalman filter will be covered in this chapter. The program has been designed to take snaps within an arc of 120 degrees at any number of intermediate points defined by the user (say n) and a regular polygon with (3n) number of sides is estimated from the given radius. Then the path is generated by trisecting the regular polygon. For example, if the number of images required is 6 and the radius as 1000 mm, then the path is defined by trisecting a regular 18 sided polygon of radius 1000 mm. The robot moves along each side of the polygon and takes snaps at each vertex point facing towards the polygon. The sample image thus obtained is used to obtain 3D information about the planar object in the robot's environment. This is achieved by using a recursive formulation of the Kalman filter discussed in chapter 15.

The program was developed for the client–server architecture, where the server program is written in C++ using the ARIA and SVS libraries, already discussed in Chap. 5 and the client is written in JAVA.

The necessary input is provided to the robot through the client program. The server then samples the images and sends them to the client along with other required information such as the robot position, the camera span angle, tilt, heights etc. In the client, 2D information of the images along with the camera parameters are used by another program called Kalman.jav to estimate the 3D information of the planar object using the Kalman filter.

16.2 Flow Chart and Source Code for 3D Perception

The client and server flow charts used for the 3D perception program are illustrated in Figs. 16.1 and 16.2 and their sample programs written in C++ and Java are given in Listings 16.1 and 16.2 respectively at the website of the book. The program output with log session is depicted in Figs. 16.3 through 16.9.

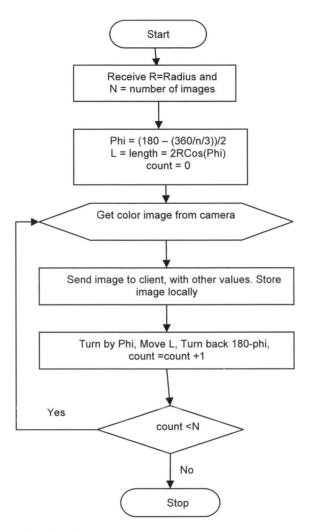

Fig. 16.1. Flowchart for client program

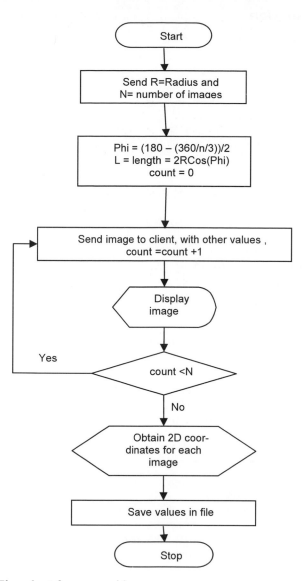

Fig. 16.2. Flowchart for server side program

Program log session

```
Opened the server port
Client has connected

 1000|6|□

1000    6


one
phi=80.00  and dist=347.30
TCP connection status = 4.
Could not connect to simulator, connecting to robot
through serial.
IEEE 1394 interface open request
1 card(s) found, 2 node(s)
Checking card 0, node 0
Vitana api addr: 78080600
Vendor length is 10
Vendor is: VITANA
Model length is 10
Model is: PixeLINK(tm)
Camera found at node 0 0: VITANA PixeLINK(tm)
Camera ISO bandwidth needed: A10
Max_Image_Size_Inq: 05080408
Unit_Size_Inq: 00080008 (0)
Image_Size_Inq: 05000400 (0)
Frame_Rate_min: 0000000E (0)
Frame_Rate_max: 000000A0 (0)
Frame_Rate_def: 00000011 (0)
Flags: 00000000 (0)
PCS2112 ver 0x30
Imager reset starting...
Imager reset succeeded
Imager ready.
Camera ISO speed set to 400 Mb/sec
Camera ISO parameters: 2000000
Opened frame grabber.
Size: 320/1280 240/960
Size: 320/1280 240/960
Starting DMA
Camera ISO bandwidth needed: A10 (2576)
w, h: 640, 480
frame bytes: 614400
Size: 196608
buf size: 196608
Iso Thread running..., dmafd = 10 (43285000 42ADA010)
```

```
DBS: 161   FN: 1   QPC: 1 MAX_DBS: 4
Syncing 0
Syncing 1
Syncing 2
Connected to robot.
Name: arcane
Type: Pioneer
Subtype: p2de
Loaded robot parameters from p2de.p
Connected
Size: 320/1280 240/960
54
SEND

RECD
Stopping DMA
ISO thread terminated
turn -80.00
move 347.30
turn 100.00
Size: 320/1280 240/960
Starting DMA
Camera ISO bandwidth needed: A10 (2576)
w, h: 640, 480
frame bytes: 614400
Size: 196608
buf size: 196608
Iso Thread running..., dmafd = 10 (43285000 42ADA010)
DBS: 161   FN: 1   QPC: 1 MAX_DBS: 4
54
SEND

RECD
Stopping DMA
ISO thread terminated
turn -80.00
move 347.30
turn 100.00
Size: 320/1280 240/960
Starting DMA
Camera ISO bandwidth needed: A10 (2576)
w, h: 640, 480
frame bytes: 614400
Size: 196608
buf size: 196608
Iso Thread running..., dmafd = 10 (43285000 42ADA010)
DBS: 161   FN: 1   QPC: 1 MAX_DBS: 4
```

```
54
SEND

RECD
Stopping DMA
ISO thread terminated
turn -80.00
move 347.30
turn 100.00
Size: 320/1280 240/960
Starting DMA
Camera ISO bandwidth needed: A10 (2576)
w, h: 640, 480
frame bytes: 614400
Size: 196608
buf size: 196608
Iso Thread running..., dmafd = 10 (43285000 42ADA010)
DBS: 161  FN: 1  QPC: 1 MAX_DBS: 4
54
SEND

RECD
Stopping DMA
ISO thread terminated
turn -80.00
move 347.30
turn 100.00
Size: 320/1280 240/960
Starting DMA
Camera ISO bandwidth needed: A10 (2576)
w, h: 640, 480
frame bytes: 614400
Size: 196608
buf size: 196608
Iso Thread running..., dmafd = 10 (43285000 42ADA010)
DBS: 161  FN: 1  QPC: 1 MAX_DBS: 4
54
SEND

RECD
Stopping DMA
ISO thread terminated
turn -80.00
move 347.30
turn 100.00
Size: 320/1280 240/960
Starting DMA
```

```
Camera ISO bandwidth needed: A10 (2576)
w, h: 640, 480
frame bytes: 614400
Size: 196608
buf size: 196608
Iso Thread running..., dmafd = 10 (43285000 42ADA010)
DBS: 161   FN: 1   QPC: 1 MAX_DBS: 4
54
SEND


RECD
Stopping DMA
ISO thread terminated
turn -80.00
move 347.30
turn 100.00
Done , exiting
Disconnecting from robot.
Lost connection
Closing video device
```

Fig. 16.3. Output of input window

Fig. 16.4. Output window depicts the selection of the first point

Fig. 16.5. Output window depicts the selection of the second point

Fig. 16.6. Output window depicts the selection of the third point

Fig. 16.7. Output window depicts the selection of the fourth point

Fig. 16.8. Output window depicts the selection of the fifth point

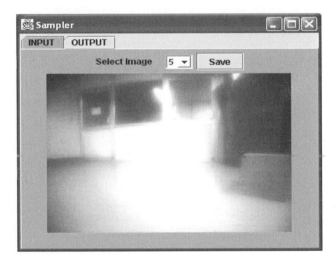

Fig. 16.9. Output window depicts the selection of the sixth point

After starting of the server program, it opens a socket and listens for a client request. When the client is connected to the server, it launches the input window where the user enters the required input, such as the radius for the robot path and the number of images to be taken. The file name by which the images will be stored in the robot must also be specified in this window, which can be seen in Fig. 16.3. When the "Submit" button is pressed the server starts the Sampler procedure and the images are sent one by one, which are depicted in Figs. 16.4 through 16.9. The output window has a drop-down list using which one may select the image number. After the images are available at the client the user can specify the fixed point of interest in each image by clicking on the image at the desired point. These 2D values along with the robot position and camera parameter received from the server will be stored in a file named input.dat when the "Save" button is pressed, which is given in Table 16.1. The values stored in input.dat are data for the images shown in Figs. 16.4 through 16.9.

Table 16.1. Values saved into input.dat

−1.8	1.7	9.0	14.0	28.5	−0.262	−1.574	0.0
−1.8	2.0	18.0	15.5	28.5	0.0	−1.574	0.0
−4.4	1.7	24.0	13.5	28.5	0.0	−1.574	0.0
−1.7	1.0	26.5	13.7	28.5	0.227	−1.574	0.0
−3.2	0.8	32.5	14.8	28.5	0.262	−1.574	0.0
−5.0	−0.5	28.0	3.0	28.5	0.0	−1.574	0.0

Table 16.2. 3D output after every iteration

After 1st iteration	9.0643356698	13.9744013198	28.49088823106
After 2nd iteration	9.5242775216	16.5935998894	29.51163005427
After 3rd iteration	10.1995493714	15.0153226290	32.4647295181
After 4th iteration	10.5980021371	11.0918751729	30.315874159
After 5th iteration	11.3524524995	10.2802239269	33.0284420210
After 6th iteration	11.8455499713	2.4803175141	33.7681237207

16.3 Summary

This chapter has described the program for taking 2D images from various viewpoints in the surroundings of the robot and storing them in a given file and later using this information for reconstruction of 3D points using the Kalman filter. The source code of the program is available in Listing 16.3 at the website of the book. The input to the program is given from `input.dat` shown in Table 16.1 and the program generates the output shown in Table 16.2.

17 Perceptions of Non-planar Surfaces

17.1 Introduction

A novel technique for automated perception of a non-planar surface is covered in this chapter. From the camera image, the area of interest is extracted first using the curve tracing method and next the nature of the curve is predicted by using a piecewise linear approximation method.

17.2 Methods of Edge Detection

An edge is a contour of pixels that separates two regions of different intensities. It can be defined as a contour along which the brightness in the image changes abruptly. A very simple method for finding edges is to evaluate the directional derivatives of $g(x, y)$ in the x- and y-directions, which is known as a gradient filter; g_1 and g_2 respectively denoted as follows:

$$g_1 = \frac{\partial g(x, y)}{\partial x} \qquad \text{and} \qquad g_2 = \frac{\partial g(x, y)}{\partial y}$$

The resulting gradient can be evaluated by the vector addition of g_1 and g_2 and is given by

$$g = \left[g_1^2 + g_2^2 \right]^{1/2} \qquad \text{and phase} \quad \phi = \tan^{-1}\left(\frac{g_1}{g_2} \right)$$

A pixel is said to lie on an edge if the gradient g is above a specified threshold. Based on this concept, various types of edge detection filters are available in the literature [Clark, 1989; Gonzalez, 1993; Heijden, 1995; Fram, 1975; Marr, 1980]. The gradient filter, compass filter and Laplace filter are a few among them. The common gradient filters such as Prewitt, Sobel and isotropic filters compute horizontal and vertical differences of local sums and reduce the effect of noise in the image data. All these fil-ters have desirable properties of yielding zeros for uniform regions.

Computer vision systems often demand the segmentation of a scene into constituent objects, which is based on the object boundaries. The object boundary is represented by the edge, which is nothing but an abrupt change in the gray levels. A spatial derivative of the image $f(x, y)$ assumes a local maximum in the direction of an edge shown in Fig. 17.1, which is used to measure the gradient of f along r in a direction θ.

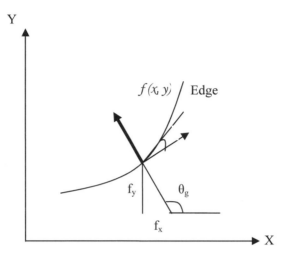

Fig. 17.1. Finding the directional derivative of the curve $f(x, y)$

$$\frac{\partial f}{\partial r} = \frac{\partial f}{\partial x} \cdot \frac{\partial x}{\partial r} + \frac{\partial f}{\partial y} \cdot \frac{\partial y}{\partial r} \tag{17.1}$$

$$= f_x \cos \theta + f_y \sin \theta$$

The maximum value of $\frac{\partial f}{\partial r}$ is obtained when $\frac{\partial}{\partial \theta}\left(\frac{\partial f}{\partial r}\right) = 0$, i.e.

$$-f_x \sin\theta + f_y \cos\theta = 0$$

$$\Rightarrow \theta_g = \tan^{-1}\left(\frac{f_y}{f_x}\right)$$

$$\left(\frac{f_y}{f_x}\right)_{max} = \sqrt{f_x^2 + f_y^2} \qquad (17.2)$$

where θ_g is the direction of the edge.

Based on these equations, there are two types of edge detection operators are there, such as gradient operators and compass operators, also called as masks. The operators represent a finite difference approximation of either the orthogonal gradients f_x, f_y or the directional gradient $\frac{\partial f}{\partial r}$. Let H denotes a $p \times p$ mask. For an arbitrary image U the inner product (H operated on U) at location (m, n) is given by the correlation

$$\langle U, H \rangle_{m,n} \triangleq \sum_i \sum_j h(i,j)u(i+m, j+n) = u(m,n) \otimes h(-m,-n) \qquad (17.3)$$

Gradient operators: There are pairs of masks H_1 and H_2, which measure the gradient of an image u (m, n) in orthogonal directions.

Let

$$g_1(m,n) \triangleq \langle U, H_1 \rangle_{m,n} \quad \text{and} \quad g_2(m,n) \triangleq \langle U, H_2 \rangle_{m,n}$$

where g_1 and g_2 are the horizontal and vertical gradient vectors, respectively.

Then the magnitude and direction of the gradient vector are given by

$$g(m,n) = \sqrt{g_1^2(m,n) + g_2^2(m,n)}$$

$$\theta_g = \tan^{-1}\frac{g_2(m,n)}{g_1(m,n)} \qquad (17.4)$$

Some common gradient operators are Prewitt's operator and Sobel's operator having the form given below:

	H_1	H_2
Prewitt's operator	$\begin{bmatrix} -1 & 0 & 1 \\ -1 & [0] & 1 \\ -1 & 0 & 1 \end{bmatrix}$	$\begin{bmatrix} -1 & -1 & -1 \\ 0 & [0] & 0 \\ 1 & 1 & 1 \end{bmatrix}$
Sobel's operator	$\begin{bmatrix} -1 & 0 & 1 \\ -2 & [0] & 2 \\ -1 & 0 & 1 \end{bmatrix}$	$\begin{bmatrix} -1 & -2 & -1 \\ 0 & [0] & 0 \\ 1 & 2 & 1 \end{bmatrix}$

The gradient map $g\,(m,\,n)$ needs to undergo a threshold operation. An edge is a point declared at the pixel point $(m,\,n)$ if $g\,(m,\,n)$ exceeds the threshold 't' which can be set by a study of the cumulative histogram of $g(m, n)$ so that 5–10% of pixels with the largest gradients are declared as edges. Therefore, the edge map $\varepsilon\,(m,\,n)$ is given as

$$\varepsilon(m,\ n) = \begin{cases} 1 & g(m,\ n) > t \\ \\ 0 & otherwise \end{cases} \tag{17.5}$$

Edge maps generated by the above operators usually have thick solid lines as boundaries which themselves may have two edges. Thinning algorithms help to transform such an image to a set of simple digital arcs, which lie roughly along the medial axes.

17.3 Curve Tracking and Curve Fitting

Edge detection operators generate an edge map which is a spatial distribution of pixels constituting edges of an image. However, interpretive analysis of the image would require knowledge of the nature of the edges which requires knowledge of the connectivity or continuity of the edge, which requires a contour following method or edge tracking. This method obtains the pixels of an edge as a continuous array, where each successive member is the neighboring point lying entirely on the edge. This can be done by using dynamic programming to find the optimum path between two given pixels on the edge. The procedure for tracking the curve from the grabbed image is given in Listing 17.1.

Listing 17.1 Procedure for tracking of the curve from the grabbed image

```
Begin:
Find image // Collect the image (Im) from the camera of
the robot and transfer  it to the client machine //
Find the pixel matrix (Pm) from the image.

If:
The image is a color image
Then:

Convert it to grayscale values.

Apply the Prewitt's operator to the pixel matrix to
calculate the horizontal (Hg) and vertical (Vg) gradient
and find their vector sum as G.

Find a suitable threshold // a suitable threshold value
(T) is chosen for the pixel matrix //

If:   G > T
Then:

It is an edge.
Else:
End for
Arrange the points obtained by the above procedure to
represent a continuous curve.
End for
End;
```

Curve fitting: For a given set of points, finding the equation of a curve of best fit for the given values is called curve fitting. Once the points defining a curve are traced from the edge map, the problem becomes one of finding an equation for the arbitrary constants, which best defines the curve. One may proceed using one of the following methods: (i) graphical method; (ii) method of least squares; (iii) method of moments and (iv) method of group averages.

When the curve representing the given data is a linear law, $y = mx + c$, the graphical method proceeds as follows:

1. Plot the given points as a graph with a suitable scale.
2. Draw the straight line of best fit such that points are equally distributed about the line.
3. Taking two arbitrary points (x_1, y_1) and (x_2, y_2) one may find m and c.

When the points are not approximated by a straight line, one may generate a smooth curve, using the law governing the curve and reduce it to linear form. The graphical method, however, fails to produce an accurate and unique fit in this case. The principle of least squares does not help much to determine the form of the approximate curve which can fit given data. It determines the best possible coefficient in the equation when the nature of the curve is known a priori. As the order of the polynomial in (17.1) increases it becomes exceedingly difficult to solve the equations. Moreover the principle is hard to implement on curves defined by quadratic (or higher) equations in both x and y (e.g. the family of curves given by $\frac{x^2}{a^2} + \frac{y^2}{b^2} = 1$, i.e. the circle, ellipse and hyperbola).

This chapter aims at finding an alternative method to predict the nature of the curve. This method is a graphical method to estimate the curvature of a set of points. The points extracted from the edges of the image should have in good proximity and sufficient continuity. Every curve has a characteristic slope changing pattern, which can be found using the tangents, at successive points in the curve. Since it is difficult to predict tangency at all instantaneous point we shall study the secants which cut chords of equal length such that the chords form a continuous set of connected line segments. Therefore we will study the pattern of a piecewise linear approximation of curve. This principle is demonstrated for the case of a circle and an ellipse in Fig. 17.2.

The curve is approximated to form a continuously connected line segment. The size of the line segment is approximated to the curve with a threshold value of slope. The piecewise linear approximation assumes that the slope of each line segment approximates the tangent to the curve at the medial point and that the curve does not change abruptly in between two such points, and such changes would be removed as noise. The change in slope from one line segment to other is represented by the angles $\theta_1, \theta_2, \ldots$ and these can be observed as a pattern characteristic of the curve. The line segments have been extended beyond the curve to form secants for ease of visual inspection.

(a) The case of a circle

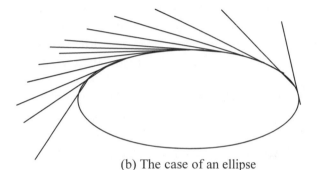

(b) The case of an ellipse

Fig. 17.2. The slope variations in a piecewise linear approximation of circular and elliptic curves

Consider an arbitrary circle in 2D space as shown in Fig. 17.2 (a). By visual inspection one can instinctively see the relation behind the changing slope of the secants. The angles formed by consecutive chords are all equal in the case of a circle, which can be proved by simple geometric axioms and theorems. Similarly for an ellipse the angles θ_1, θ_2 shown, gradually decrease to a minimum and then increase gradually, thus forming a periodically changing angle. It is, however, to be noted that to obtain a fairly accurate relation, the chord length should be chosen judiciously. Too small chord lengths would cause a number of anomalous results arising due to noise present in the edge-mapped images. Too large values reduce the accuracy and efficiency of the prediction. This depends on the nature of the algorithm being used. Once the nature of the curve is predicted, its parameters such as the center, radius, foci, etc. can be estimated by applying the principles of coordinate geometry.

In this experimental setup, images sent by a digital camera mounted on a mobile robot are analyzed and interpreted to predict the shape of obstacles and objects in the path. Edge detection has been implemented using the Prewitt gradient operator with a heuristic threshold decision. The edge map is then transformed to a simple arc diagram using the thinning algorithm and the thin curves are then tracked out to obtain an array of points which are analyzed to predict the nature of the curve. The pseudo-code for the program is given in Listing 17.2.

Listing 17.2 Procedure for Curve Fitting

```
Begin:
Divide the curve into small numbers of segments
Find gradient // obtain the slope of the tangents (θᵢ)
at each point of the segment on the curve //
If:
All the slopes of the segment are nearly equal
Then:
Predict the curve as a circle
ElseIF:
The slopes gradually decreases first to a minimum and
increases gradually
Then:
Predict the curve as an ellipse
End;
```

17.4 Program for Curve Detector

The curve detection program is aimed at identifying the curved edges of objects in the robot's environment. The program is basically divided into two basic sections, i.e. edge map generator and curve recognition system. The edge map generator is a conventional algorithm which implements a Prewitt or Sobel edge detection operator on an image. The "image" in this case is procured from a color image server program that runs on the robot and sends frames to the Java client. The rest of the processing is done on the client for better efficiency. After the generation of the edge map, a portion of the edge map is selected on the basis of its content and passed to the curve recognition system. Curve detection is based on a piecewise linear approximation theory. This theory based on the fact that every curve exhibits a distinct slope pattern which can be studied from the differential angle formed by secants of equal length cut successively through the curve.

The flow charts for the server and client program are illustrated in Figs. 17.3 and 17.4 respectively and the source code for the server and the client are given in Listing 17.3 and Listing 17.4 respectively at the website of the book. The program log session is given below followed by program output shown in Figs. 17.5 and 17.6.

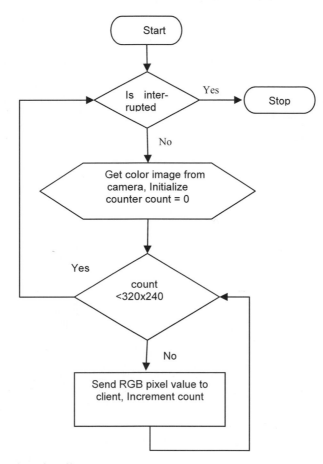

Fig. 17.3. Flow chart for client program

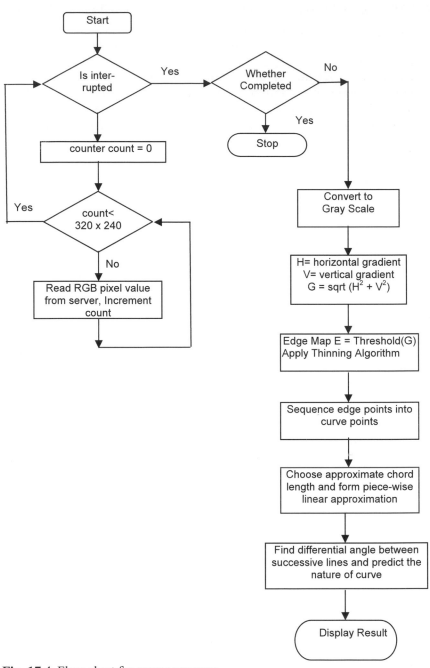

Fig. 17.4. Flow chart for server program

Program log session

```
IEEE 1394 interface open request
1 card(s) found, 2 node(s)
Checking card 0, node 0
Vitana api addr: 78080600
Vendor length is 10
Vendor is: VITANA
Model length is 10
Model is: PixeLINK(tm)
Camera found at node 0 0: VITANA PixeLINK(tm)
Camera ISO bandwidth needed: A10
Max_Image_Size_Inq: 05080408
Unit_Size_Inq: 00080008 (0)
Image_Size_Inq: 050003C0 (0)
Frame_Rate_min: 0000000E (0)
Frame_Rate_max: 000000A0 (0)
Frame_Rate_def: 00000020 (0)
Flags: 00000000 (0)
PCS2112 ver 0x30
Imager reset starting...
Imager reset succeeded
Imager ready.
Camera ISO speed set to 400 Mb/sec
Camera ISO parameters: 2000000
Opened frame grabber.
Size: 320/1280 240/960
Opened the server port
Client has connected
Size: 320/1280 240/960
Starting DMA
Camera ISO bandwidth needed: A10 (2576)
w, h: 640, 480
frame bytes: 614400
Size: 196608
buf size: 196608
Iso Thread running..., dmafd = 4 (431D7000 42A16010)
DBS: 161  FN: 1  QPC: 1 MAX_DBS: 4
DBC count error... 220 254 3 165512
line count error 0 3 114124 0
DBC count error... 237 12 3 152560
DBC count error... 3 37 3 170560
DBC count error... 168 202 3 188460
DBC count error... 164 198 3 73320
DBC count error... 24 58 2 160556
DBC count error... 36 105 3 177644
DBC count error... 226 4 2 164020
```

```
DBC count error... 26 57 3 169140
DBC count error... 84 118 3 12084
DBC count error... 194 200 2 154276
DBC count error... 213 248 3 156604
DBC count error... 130 151 2 141584
DBC count error... 189 224 3 19172
DBC count error... 157 191 3 37408
DBC count error... 230 8 3 108764
DBC count error... 9 43 3 126064
DBC count error... 69 88 3 8876
DBC count error... 71 105 3 152936
DBC count error... 227 0 3 6060
DBC count error... 109 144 3 127508
DBC count error... 202 239 2 65512
DBC count error... 140 170 2 127644
DBC count error... 200 234 4 12728
DBC count error... 80 114 3 132560
DBC count error... 199 218 3 45760
DBC count error... 236 11 2 103376
DBC count error... 20 55 4 125200
DBC count error... 186 223 2 103784
Aria: Received signal 'SIGINT'. Shutting down.
Closing video device and TCP connection ...
Stopping DMA
Closing video device and TCP connection ...
Stopping DMA
ISO thread terminated
```

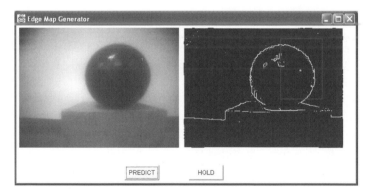

Fig. 17.5. Java front-end for edge map generator

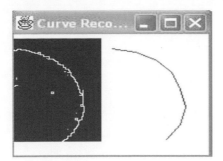

Fig. 17.6. Java frame for curve detector

During the program execution, the client front end displays the output as shown in the first window of Fig. 17.5. This image is grabbed by the robot's camera and the second window displays the edge map, as generated by the program. The window provides two control buttons namely "HOLD" and "PREDICT". The "HOLD" button serves to freeze the video frame so that a steady picture may be used for further processing. A portion of the image is then selected on the edge-map using the mouse. The "PREDICT" button launches the curve recognition program for the selected portion and the result is displayed as shown in Fig. 17.6. The gray lines represent a possible circular region with the break lines showing deviation. The effect of noisy image however is a limitation to the performance of the program.

17.5 Summary

The task of curve fitting becomes exceedingly difficult and error prone in real-time visual feedback robotic systems. Thus predictions based on traditional curve fitting methods do not provide satisfactory accuracy in a noisy visual environment. It is therefore required to approximate a certain length of the obtained curve points to a known law so that the effects of noise can be minimized. Such curve-fitting algorithms, however, lead to tedious calculations when the nature of the curve is of second or higher order. In this experiment, an alternative to traditional curve-fitting algorithms is proposed by assuming the nature of the curve as being piecewise linear. The experiment has produced satisfactory results. The effect of a noisy image, however, is a limitation on the performance of the program.

18 Intelligent Garbage Collection

18.1 Introduction

Intelligent garbage collection is one of the real-time applications of mobile robots, it utilizes the robot's motion, its gripper, as well as its vision systems. The objective of this program is to collect a given set of colored boxes and bring them to one place. The color of the box is not a restriction, and is set during the run time of the program. This is an interesting program, where the robot collects the objects having definite visual characteristics.

First the color of the box is selected and the images are obtained. Each image is checked pixel by pixel with respect to the color of the box. If the pixel value lies in a particular range (the range being set by the user) then that pixel is made black otherwise white. In this way the whole image is mapped into another image, where we have black dots resembling the color of the box. After this the image is divided into 12 blocks (3 × 4). The number of dots in each block is counted and the block having the highest number is found. If this value exceeds the preset value (the preset value is set so that noise is not taken as the object) then that particular block is selected. The robot moves to the object for grabbing. If the surface plane of the object is not parallel to the gripper arm, then it may not be possible for the robot to grab in the first attempt and some more trials are required.

18.2 Algorithms and Source Code for Garbage Collection

The client–server algorithms are given in Listings 18.1 and 18.2 and their programs written in C++ and Java are given in Listings 18.3 and 18.4 respectively at the website of the book. The garbage collection program output with program log session is depicted in Fig. 18.1.

Listing 18.1. Server algorithm

Step 1:	Open the server socket and wait for the client to join.
Step 2:	Send the image and wait for the next instruction.
Step 3:	If next instruction = IMAGE
	Then go to Step – 2.
	Else go to Step – 4.
Step 4:	If next instruction = STOP
	Then stop the robot.
	Go to Step – 2.
	Else go to Step – 5.
Step 5:	If next instruction = EXIT
	Then go to Step – 11.
	Else go to Step – 6.
Step 6:	If next instruction = GO
	Then go to Step – 7.
	Else go to Step – 2.
Step 7:	Send Image and get angle and instruction.
Step 8:	If angle = 45
	Then turn by angle. Go to Step – 9.
	Else turn by angle.
	Move 250 mm.
	Goto Step – 9.
Step 9:	If Beam break state = true
	Then close the gripper.
	Turn towards origin and move to origin.
	Drop the object there and return to initial position.
Step 10:	If instruction = IMAGE
	Go to Step – 7.
	Else If instruction = STOP Go to Step – 2.
	Else Go to Step – 11.
Step 11:	If Client connected.
	Then Go to Step – 7.
	Else Go to Step – 12.
Step 12:	Stop.

Listing 18.2. Client algorithm

Step 1:	Open the client socket and connect to the server.
Step 2:	Get the image.
Step 3:	Select the color of the box. Get the user instruction.
Step 4:	If instruction = IMAGE
	Go to Step – 2.
	Else if instruction = GO
	Go to Step – 5.
	Else go to Step – 2.
Step 5:	Get the next image.
Step 6:	I=0, C=color of the box,
	Threshold = value of range, M = min. upper limit.
	Image() = color image, Ang(12) = the angles for 12 blocks
Step 7:	For I=0 , I<320×240 , I++
	If abs(image(I)-C) <= threshold
	Image(I) = black
	k = (i/320)/80*4 + (i%320)/80;
	count[k] ++
	Else Image(I) = white
Step 8:	If MAX(count) > M
	Then keep k such that count(k) = MAX(count)
Step 9:	Send Ang(k) if k has max count
	Else send 45.
Step 10:	If Client connected
	Goto Step – 5.
	Else goto Step – 11.
Step 11:	Stop.

Program output log

```
IEEE 1394 interface open request
1 card(s) found, 2 node(s)
Checking card 0, node 0
Vitana api addr: 78080600
Vendor length is 10
Vendor is: VITANA
Model length is 10
Model is: PixeLINK(tm)
```

```
Camera found at node 0 0: VITANA PixeLINK(tm)
Camera ISO bandwidth needed: A10
Max_Image_Size_Inq: 05080408
Unit_Size_Inq: 00080008 (0)
Image_Size_Inq: 05000400 (0)
Frame_Rate_min: 0000000E (0)
Frame_Rate_max: 000000A0 (0)
Frame_Rate_def: 00000008 (0)
Flags: 00000000 (0)
PCS2112 ver 0x30
Imager reset starting...
Imager reset succeeded
Imager ready.
Camera ISO speed set to 400 Mb/sec
Camera ISO parameters: 2000000
Opened frame grabber.
Size: 320/1280 240/960
Opened the server port
Client has connected
Size: 320/1280 240/960
Starting DMA
Camera ISO bandwidth needed: A10 (2576)
w, h: 640, 480
frame bytes: 614400
Size: 196608
buf size: 196608
Iso Thread running..., dmafd = 4 (431D7000 42A16010)
DBS: 161   FN: 1   QPC: 1 MAX_DBS: 4
DBC count error... 108 142 3 150084
DBC count error... 12 42 2 5148
DBC count error... 52 62 4 83272
DBC count error... 228 244 3 47500
DBC count error... 35 69 3 120996
DBC count error... 91 129 3 106072
DBC count error... 104 130 4 124016
DBC count error... 138 175 2 116968
DBC count error... 235 17 3 121468
DBC count error... 164 195 3 137528
DBC count error... 87 121 3 115876
DBC count error... 234 0 3 57524
DBC count error... 83 118 3 36856
DBC count error... 178 206 3 156204
DBC count error... 47 66 3 82984
DBC count error... 73 107 3 112568
DBC count error... 18 46 3 158060
DBC count error... 51 86 3 90124
DBC count error... 117 152 3 137812
```

```
DBC count error... 32 66 2 161312
DBC count error... 192 226 3 118320
Got the object.
Dropping the object.
Closing video device and TCP connection ...
Stopping DMA
Aria: Received signal 'SIGINT'. Shutting down.
Closing video device and TCP connection ...
Stopping DMA
ISO thread terminated
```

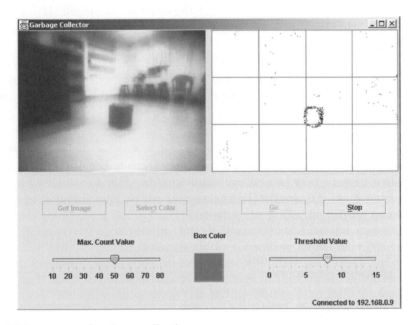

Fig. 18.1. Output of garbage collection program

18.3 Summary

In the main menu, there are four buttons, namely "Get Image", "Color", "Go" and "Stop". The function of the first button Get Image is to get the image from the server usually to take the image of the colored box. This is used to set the color of the box during the run time of the program. After the image of the box is received, which will be shown in the left window, the Color button is pressed. After that, all other buttons will be deactivated and, therefore, the portion of the image that contains the color of the box is selected. Then, the color of the box taken by the program is shown in the Box Color region. If the color of the box is not taken correctly or we want to select the color of the box again, the Get Image button is clicked again, and we proceed as before.

Once the user is satisfied with the color of the box, then the Go button is clicked. This makes the robot search for the box in its environment. If it finds one, then it moves towards it until it grabs it and it drops it at the origin. After dropping at the origin it moves back to its original position from where the box was collected. Again, it continues to scan the environment. This program runs continuously until the window is closed. If for some reasons or other the user wants to stop the robot, the Stop button is clicked, which stops the robot in its next cycle of getting the image.

The program has two sliders, i.e. Threshold value and Max. count value, for checking whether the image contains the color box or not. The program finds out which pixels are in close range of color as compared to the color of the box. This range is set by the Threshold value slider. Once the color box is found in the image, the user has to decide in which direction the robot should move in order to catch the object. For which the right window is divided into 12 parts (3 × 4) and we count the number of pixels close to the color of the Color box in each sub-window. The sub-window having the largest count decides the direction of the robot. This count is set by the Max. count value slider.

References

[Aloimonos,1987] Aloimonos, J., Weiss, I., and Bandopadhay, A., "Active vision", *International Journal of Computer Vision*, vol. 1, no. 4, pp. 333–356, 1987.

[Arbib, 1981] Arbib, M., "Perceptual structures and distributed motor control", in *Handbook of Physiology—The Nervous System II,* ed. V.B. Brooks, Bethesda, Maryland: American Physiological Society, pp. 1449–1465, 1981.

[Asada, 1990] Asada, M., "Map Building for a mobile robot from sensory data", *IEEE Transaction on Systems, Man and Cybernetics*, vol. 37, no. 6, pp. 1326–1336, Nov./Dec., 1990.

[Ayache, 1987] Ayache, N., Faugeras, O.D., "Building a consistent 3D representation of the mobile robot environment by combining multi stereo views", *Proceedings of the International Joint Conference on Artificial Intelligence*, August 1987.

[Ayache, 1991] Ayache, N., *Artificial Vision for Mobile Robot*, The MIT Press, Massachusetts, 1991.

[Baldi, 1995] Baldi, P.F., Hornik, K., "Learning in linear neural networks: a survey", *IEEE Transactions on Neural Networks*, vol. 6. no. 4, pp. 837–857, July 1995.

[Ballard, 1991] Ballard, D.H., "Animate vision", *Journal of Artificial Intelligence*, vol. 48, no. 1, pp. 57–86, 1991.

[Bezdek, 1991] Bezdek, J.C., Ed., *Pattern Recognition with Fuzzy Objective Function Algorithms*, Kluwer Academic Press, 1991.

[Bharick, 1984] Bharick, H.P., "Semantic memory content in permastore: fifty years of memory for Spanish learned in school", *Journal of Experimental Psychology: General*, Vol. 120, pp. 20–33, 1984.

[Borenstain, 1996] Borenstain, J., Everett, H.R. and Fang, L. *Navigating Mobile Robots: Systems and Techniques*, A. K. Peter Wellesley, 1996.

[Briscoe, 1997] Briscoe, G., Caelli, T., "ABC: biologically motivated image understanding", in *Machine Learning and Image Interpretation*, editor, Caelli, T., and Bischof, W.F., Plenum Press, New York, 1997.

[Brooks, 1989] Brooks, R.A., "A robot that walks; emergent behaviors from a carefully evolved network", *Neural Computation* 1(2):253, 1989.

[Brown, 1997] Robert, G., Hwang, Patrick Y.C., *Introduction to Random Signals and Applied Kalman Filtering*, John Wiley, 1997.

[Buchanan, 1993] Buchanan, B.G. and Wilkins, D.C., Eds., *Readings in Knowledge Acquisition and Learning: Automating the Construction and Improvement of Expert Systems*, Morgan Kaufmann, San Mateo, CA, 1993.

[Caelli and Bischob, 1997] Caelli, T., and Bischob, W.F., *Machine Learning and Image Interpretation*, Plenum Press, New York, 1997.

[Carpenter, 1987] Carpenter, G.A., Grossberg, S., "A massively parallel architecture for a self-organizing neural pattern recognition machine", *Computer Vision Graphics and Image Processing* 37, pp. 54–115, 1987.

[Chang, 1986] Chang, T.M., "Semantic Memory: Facts and Models", *Psychological Bulletin*, vol. 99, pp. 199–220, 1986.

[Clark, 1989] Clark, J.J, "Authenticating edges produced by zero-crossing algorithms", *IEEE Transactions on Pattern Analysis and Machine Intelligence*, vol. 12, no. 8, pp. 830–831, 1989.

[Connell, 1990] Connell, J., *Minimalist Mobile Robotics: A Colony-Style Architecture for an Artificial Creature*, Boston, Mass: Academic Press, 1990.

[Davis, 1986] Davis, L.S., and Kambhampati, S., "Multi-resolution path planning for mobile robots", *IEEE Transactions on Robotics and automation*, vol. RA-2, no. 3, 1986.

[Dean, 1995] Dean, T., Allen, J., and Aloimonos, Y., *Artificial Intelligence: Theory and Practice*, Addison-Wesley, 1995.

[Elfes, 1987] Elfes, A., "Sonar-based real-world mapping and navigation", *IEEE Journal of Robotics and Automation*, vol. RA-3, no. 3, pp. 249–264, June, 1987.

[Feldman, 1992] Feldman, J.A., "Natural Computation and Artificial Intelligence", Plenary Lecture presented at the International Joint Conference on Neural Networks, Baltimore, 1992.

[Fermuller and Aloimonos, 1993] Fermuller, C., and Aloimonos, Y., "Vision and action", *Image Vision Computation*, vol. 13, no.10, pp. 725–744, Dec. 1993.

[Filho, 1994] Filho, R., "Genetic algorithms programming environments", *IEEE Transactions on Computers*, vol. 26, pp. 28–43, June 1994.

[Fram, 1975] Fram, J.R. and Deutsch, E.S., "On the quantitative evaluation of edge detection schemes and their comparison with human performance", *IEEE Transactions on Computers*, vol. C-24, no. 6, pp. 616–628, 1975.

[Gardner, 1985] Gardner, H., *The Mind's New Science: A History of the Cognitive Revolution*, New York: Basic Book, 1985.

[Goldberg, 1989] Goldberg, D.E., *Genetic Algorithm in Search Optimization and Machine Learning*, Reading, MA, Addison-Wesley, 1989.

[Harlick, 1993] Harlick, R.M., Shapiro, L.G., "Computer and Robot Vision", Addison-Wesley, Vol. 2, 1993.

[Hashimoto, 1999] Hashimoto, K., "Observer-based visual servoing", in *Control in Robotics and Automation*, edited by Ghosh, B.K., Xi, Xing, Tarn, J.J., Academic Press, 1999.

[Haykins, 1999] Haykin, S., *Neural Networks: A Comprehensive Foundation*, Prentice Hall, 1999.

[Heijden, 1995] Heijden F.V., "Edge and Line Feature Extraction Based on Covariance Models", *IEEE Trans. on Pattern Analysis and Machine Intelligence*, vol. 17, no. 1, pp. 69–77, Jan. 1995.

[Hinton, 1981] Hinton, G.E., "Shape representation in parallel systems", *Proceedings of the 7th International Joint Conference on Artificial Intelligence*, Vancouver, British Columbia, 1981.

[Horswill, 1993] Horswill, I., "Polly: A vision-based artificial agent", *Proceedings of the Eleventh National Conference on Artificial Intelligence*, Menlo Park, AAAI Press, CA, 1993.

[Jahne, 1997] Jahne, B., *Practical Handbook on Image Processing for Scientific Applications*, CRC Press, Boca Raton, New York, 1997.

[Jain, 1999] Jain, L. C., Fukuda, T., *Soft Computing for Intelligent Robotic Systems*, Springer-Verlag, 1999.

[Java, 2002] Java™ 2 SDK, Standard Edition Documentation (Java 2 platform API) from java.sun.com, 2002

[Kasabov, 1998] Kasabov, N., "Introduction: Hybrid intelligent adaptive systems", *International Journal of Intelligent System*, vol. 6, pp. 453–454, 1998.

[Klander, 2000] Klander, Lars, Core Visual C++ 6.0, Pearson Education, 2000.

[Konar, 2000] Konar A., *Artificial Intelligence & Soft Computing: Behavioral and Cognitive Modeling of the Human Brain*, CRC Press, Boca Raton, 2000.

[Konolige et al., 1995] Konolige, K., Myers K., Saffiotti, A., and Ruspini, E., "The Saphira architecture: A design for autonomy", *Journal of Experimental and Theoretical Artificial Intelligence*, vol. 9, pp.215–235, 1995.

[Kosko, 1987] Kosko, B. "Adaptive bi-directional associative memories", *Applied Optics*, 26 (23): pp.4947–4960, Dec. 1987.

[Kosko, 1988] Kosko, B. "Bidectional associative memories", *IEEE Transactions on Systems, Man and Cybernetics*, SMC-18, pp. 42–60, 1988.

[Kosko, 1994] Kosko, B., *Neural Networks and Fuzzy Systems*, Prentice Hall, Englewood Cliffs, NJ, 1994.

[Lee et al., 1997] Lee, Seong-Whan, and Song, Hee-Heon, "A new recurrent neural-network architecture for visual pattern recognition", *IEEE Transactions on Neural Networks*, vol. 8, no.2, pp.331–340, March 1997.

[Lin et al., 1994] Lin, H. S., Xiao, J., and Michalewicz, Z., "Evolutionary navigator for a mobile robot", *Proceedings of IEEE International Conference on Robotics and Automation*, San Diego, CA, pp. 2199–2204, May 1994.

[Maes and Brooks, 1990] Maes, P., Brooks, R., "Learning to coordinate behaviors", *Proceedings of Eighth National Conference on Artificial Intelligence*, 796, AAAI Press, Menlo Park, CA, 1990.

[Mamdani, 1977] Mamdani, E.H., *Application of fuzzy set theory to control systems, in Fuzzy Automata and Decision Processes, Gupta*, M.M., Saridies, G.N. and Gaines, B.R., Eds., Oxford University Press, Amsterdam, New York, pp. 77–88, 1977.

[Marr, 1980] Marr, D.C. and Hildreth, E.C., "Theory of edge detection", Proceedings of Royal Society, London, vol.B207, pp. 187–212, 1980.

[Marr, 1982] Marr, D., *Vision*, Freeman, San Francisco, 1982.

[Mataric, 1992] Matric, M. J., "Integration of representation into goal-driven behavior-based robots", *IEEE Transactions on Robotics and Automation,* vol.8, no.3, pp.304–312, 1992.

[Matlin, 1984] Matlin, M.W., *Cognition*, Harcourt Brace Publishers & Prism Books, 1995.

[McDermott et al., 1984] McDermott, D. and Davis, E., "Planning routes through uncertain territory", *Journal of Artificial Intelligence*, vol. 22, 1984.

[Merlo et al., 1987] Merlo, X., Lanusse, A., Zavidovique B., "Optimal control of a robot perception system", *Proceedings of IASTED International Symposium on Expert Systems Theory and Applications*, Geneve, 1987.

[Michalewicz, 1996] Michalewicz, Z., *Genetic Algorithms + Data Structures = Evolution Programs*, 3rd edition, New York, Springer-Verlag, 1996.

[Michalski, 1983] Michalski, R.S., "A theory and methodology of inductive learning", *Artificial Intelligence*, vol. 20, no. 2, pp. 111–161, 1983.

[Minsky, 1975] Minsky, M., "A framework for representing knowledge", *The Psychology of Computer Vision,* edited by Patrick H. Winston, McGraw-Hill, New York, 1975.

[Mitchell, 1997] Mitchell, T.M., *Machine Learning*, Tata McGraw-Hill, 1997.

[Mitchell et al., 2001] Mitchell, Mark, Oldham, Jeffrey and Samuel, Alex, *Advanced Linux Programming*, New Riders Publisher, 2001.

[Murphy et al., 1998] Murphy, R., Kortenkamp, D., "Vision for mobile robots", *Artificial Intelligence and Mobile Robots*, AAAI Press/The MIT Press, 1998.

[Murphy, 1998] Murphy, R.R., *Artificial Intelligence and Mobile Robots*, AAAI Press/The MIT Press, 1998.

[Narendra et al., 1990] Narendra, K. S. and Parthasarathi, K., "Identification and control of dynamic system using neural networks", *IEEE Transactions on Neural Networks*, vol.1, pp. 4–27, 1990.

[Newell et al., 1972] Newell, A., and Simon, H.A., *Human Problem Solving,* Englewood Cliffs, NJ: Prentice Hall, 1972.

[Pagac et al., 1998] Pagac, D., Nebot, E.M., and Durrant-Whyte, H., "An evidential approach to map-building for autonomous vehicles", *IEEE Transactions on Robotics and Automation*, vol. 14, no. 4, pp. 623–629, August, 1998.

[Patnaik et al., 1998a] Patnaik, S., Konar, A. and Mandal, A. K., "Map building and navigation by a robotic manipulator", *Proceedings of International Conference on Information Technology*, Tata McGraw-Hill Publisher, New Delhi, pp. 227–232, 1998.

[Patnaik et al., 1998b] Patnaik, S., Konar, A. and Mandal, A.K., "Navigational planning with dynamic scenes with timed Petri nets", *Proceedings of International Conference* on Computer and Devices for Communication, Allied Publisher, New Delhi, pp. 40–43, 1998.

[Patnaik et al., 1999a] Patnaik, S., Konar, A. and Mandal, A.K. "Constrained hierarchical path planning of a robot by employing neural nets", *Proceedings of the Fourth International Symposium on Artificial Life and Robotics*, pp. 690–693, Japan, Jan. 1999.

[Patnaik, 1999b] Patnaik, S., *Building Cognition for Mobile Robot*, Ph.D. Thesis submitted to Jadavpur University, Calcutta, India, Sept., 1999.

[Patnaik et al., 1999c] Patnaik, S., Konar, A., and Mandal, A. K., "Bi-directional associative memory for mobile robot navigation", *Proceedings of International Conference on Neural Network*, Washington, July, 1999.

[Patnaik et al., 2003a] Patnaik, S., Karibasappa, K., "Cognition techniques and their applications", *Technology and Business for the New Millennium*, edited by Prof. C.T. Leondes, Kluwer Academic Press, 2003.

[Patnaik et al., 2003b] Patnaik, S., Karibasappa, K., "Edge, shade and mixed range detection by fuzzy gaussian filter for an autonomous robot", *Journal of Intelligent Robotic Systems*, vol.37, no.3, 2003.

[Pedrycz, 1995] Pedrycz, W., *Fuzzy Sets Engineering*, CRC Press, Boca Raton, FL, 1995.

[Pomerleau et al., 1989] Pomerleau, D.A., *ALVINN: An autonomous land vehicle in a neural network*, Pittsburgh, PA: Carnegie Mellon University, Technical report CMU-CS-89-107, 1989.

[Popovic et al., 1994] Popovic, D., Bhatkar, V.P., *Methods and Tools for Applied Artificial Intelligence*, Marcel Dekker Inc, 1994.

[Gonzalez, 1993] Gonzalez, R.C., Richard E.W., *Digital Image Processing*, Addison Wesley, 1993.

[Rich et al., 1996] Rich, E., Knight, K., *Artificial Intelligence*, McGraw-Hill, New York, 1996.

[Rimon, 1992] Rimon, E., Koditschek, Daniel E., "Exact robot navigation using artificial potential functions", *IEEE Transactions on Robotics and Automation*, vol. 8, no. 5, pp. 501–518, October 1992.

[Rubini, 1998] Rubini, A., *Linux Device Drivers*, O'Reilly, 1998.

[Rumelhart et al., 1986] Rumelhart, D.E., and McClelland, J.L., *Parallel Distribution Processing: Exploration in the Microstructure of Cognition*, vol. 1, Cambridge, MA: MIT Press, 1986.

[Saffioti, 1997] Saffiotti, A., "Fuzzy logic in autonomous robotics: behavior coordination", in *Proceedings of 6th IEEE International Conference on Fuzzy Systems*, Barcelona, Spain, 1997, vol. 1, pp. 573–578.

[Samet, 1982] Samet, H., "Neighbor finding techniques for image representation by Quadtree", Journal of *Computer Graphics and Image Processing*, Vol. 18, pp 37–57, 1982.

[Saphira, 1999] *Saphira Operations and Programming Manual*, Version 6.2, August 1999.

[Schalkoff, 1997] Schalkoff, R.J., Artificial Neural Networks, McGraw-Hill, 1997.

[Sekuler et al., 1990] Sekuler, R. and Blake, R., *Perception* (Second edition), McGraw-Hill, 1990.

[Sullivan, 1999] Sullivan, M.J., Vision active deformable modes in visual servoing: *Control in Robotics and Automation*, edited by Ghosh, B.K., Xi, Xing, Tarn, J.J., Academic Press, 1999.

[Swan, 2000] Swan, T., *GNU C++ for Linux*, QUE Corporation, 2000.

[Sympson, 1988] Sympson, P., *Artificial Neural Nets: Concepts, Paradigms and Applications*, Pergamon Press, Oxford, 1988.

[Takahashi, 1989] Takahashi, O., and Schilling, R. J., "Motion planning in a plane using generalized Voronoi diagrams", IEEE *Transactions on Robotics and Automation*, vol.5, no.2, 1989.

[Tanaka, 1995] Tanaka, K., "Stability and stabilizability of fuzzy-neural-liner control systems", *IEEE Transactions on Fuzzy Systems*, vol. 3, no. 4, 1995.

[Taylor and Kriegman, 1998] Taylor, C.J., and Kriegman D.J., "Vision-Based Motion Planning and Exploration Algorithm for Mobile Robots", *IEEE Trans on Robotics and Automation*, vol. 14, no. 3, pp. 417-426, June, 1998.

[Trojanowski, 1997] Trojanowski, K., Michalewicz, Z., and Xiao, J., "Adding memory to an evolutionary planner/navigator", *Proceedings of the Fourth IEEE International Conference on Evolutionary Computation*, Indianapolis, IN, April, 1997.

[Tulving, 1987] Tulving, E., "Multiple memory systems and consciousness", *Human Neurobiology*, vol.6, pp.67–80, 1987.

[Tzionas et al., 1997] Tzionas, Panagiotis G., et al. "Collision-free path planning for a diamond-shaped robot using two-dimensional cellular automata", *IEEE Trans. on Robotics and Automation*, vol.13, no.2, 1997.

[Waltz, 1997] Waltz, D., "Neural nets and AI: time for a synthesis", *Plenary talk, International Joint Conference on Neural Networks*, Houston, vol.1, 1997.

[Winston, 1975] Winston, P.H., "Learning structural descriptions from examples", *The Psychology of Computer Vision* edited by P.H. Winston, McGraw-Hill, New York, 1975.

[Xiao et al., 1997] Xiao, J., Michalewicz, Z., Zhang L., Trojanowski K., "Adaptive evolutionary planner/navigator for mobile robots", *IEEE Transactions on Evolutionary Computation*, vol. 1, no. 1, pp. 18–28, April 1997.

[Yager, 1983] Yager, R.R., "Some relationships between possibility, truth and certainty", *Fuzzy Sets and Systems*, Elsevier, North Holland, vol. 11, pp. 151–156, 1983.

[Zadeh, 1983] Zadeh, L.A., "The role of fuzzy logic in the management of uncertainty in Expert systems", *Fuzzy Sets and Systems*, Elsevier, North-Holland, vol. 11, pp. 199–227, 1983.

[Zavidovique, 2002] Zavidovique, B.Y., "First steps of robotic perception: the turning point of the 1990s", *Proceedings of the IEEE*, vol. 90, no. 7, pp.1094–1112, July 2002.

[Zimmerman, 1991] Zimmerman H.J., *Fuzzy Set Theory and Its Applications*, Kluwer Academic, Dordrecht, The Netherlands, 1991.

Index

2D World map 21–23, 202
3D Line reconstruction 243
3D Perception 227, 251
3D Plane reconstruction 247, 248
3D Points reconstruction 237,
    241–243, 247

ARIA 79, 80
    client-server 80, 96
    socket programming 95, 96
ArRobot 80–82, 84, 85, 87, 88–91,
    93, 117, 129, 140, 153, 165, 177

BotSpeak program 127

Cellular automata 39, 40
Chromosome encoding 61
Computational theory of Marr 13
Correspondence problem 249
Crossover 20, 59, 61–66
Curve fitting 266, 267
Curve tracking 266

Edge detection 263, 265, 266, 270
Elitism 63

GA-based navigation 67
Garbage collection 277
Genetic algorithms 17, 19, 20, 60
Global coordinate system 212, 217
Global representation 204, 211,
    212, 217
Gripper control 137, 139, 144

Image capture 79, 100, 101, 221
Image formation 203, 205
Image-server program 189–190, 270
Imaging geometry 201, 205
Intelligent garbage collection 277

Kalman filter 204, 227, 228–231,
    237, 239–242, 244–249,
    251, 235

Map building 1, 21–23, 25–27, 29,
    31–33
Minimal representation 227
Motion-server program 189–190, 196
Mutation 20, 59–66

Navigation 6, 7, 10, 15, 21, 22, 39, 41,
    59, 64, 66, 67, 69, 70, 78–80, 83,
    94, 151, 189, 201
Navigator client program 189,
    195, 196

Path planning 2, 39, 41, 47, 53, 201
Perceptions
    non-planar surfaces 263
Perspective projection 203–211, 218

Quadtree 41–44, 47, 49, 52–54
    neighbor finding algorithms 47

Range devices 87, 93
Re-planning 59, 68–70
Run length encoding 190, 221

Selection 63, 66
Small vision system 79, 100
Socket programming 95, 96
Sonar reading display 151, 153, 155
SVS C++ classes 101

Tele-operation program 177, 180, 181
Temporal associative memory 68,
    69, 76

Wandering program 163

Cognitive Technologies

S. K. Pal, L. Polkowski, A. Skowron (Eds.):
Rough-Neural Computing.
Techniques for Computing with Words.
XXV, 734 pages. 2004

H. Prendinger, M. Ishizuka (Eds.):
Life-Like Characters.
Tools, Affective Functions, and Applications.
IX, 477 pages. 2004

H. Helbig:
**Knowledge Representation and
the Semantics of Natural Language.**
XVIII, 646 pages. 2006

P. M. Nugues:
**An Introduction to Language Processing
with Perl and Prolog.**
An Outline of Theories, Implementation,
and Application with Special Consideration
of English, French, and German.
XX, 513 pages. 2006

W. Wahlster (Ed.):
SmartKom: Foundations of Multimodal Dialogue Systems.
XVIII, 644 pages. 2006

B. Goertzel, C. Pennachin (Eds.):
Artificial General Intelligence.
XVI, 509 pages. 2007

O. Stock, M. Zancanaro (Eds.):
PEACH – Intelligent Interfaces for Museum Visits.
XVIII, 316 pages. 2007

V. Torra, Y. Narukawa:
**Modeling Decisions: Information Fusion
and Aggregation Operators**
XIV, 284 pages. 2007

P. Manoonpong:
**Neural Preprocessing and Control
of Reactive Walking Machines.**
Towards Versatile Artificial Perception–Action Systems.
XVI, 185 pages. 2007

S. Patnaik
Robot Cognition and Navigation
An Experiment with Mobile Robots
XVI, 290 pages. 2007

Printing: Krips bv, Meppel
Binding: Stürtz, Würzburg